SpringerBriefs in Electrical and Computer Engineering

Computational Electromagnetics

Series editor

Rakesh Mohan Jha, Bangalore, India

More information about this series at http://www.springer.com/series/13885

Balamati Choudhury · Bhavani Danana
Rakesh Mohan Jha

PBG based Terahertz Antenna for Aerospace Applications

 Springer

Balamati Choudhury
Centre for Electromagnetics
CSIR-National Aerospace Laboratories
Bangalore, Karnataka
India

Rakesh Mohan Jha
Centre for Electromagnetics
CSIR-National Aerospace Laboratories
Bangalore, Karnataka
India

Bhavani Danana
Centre for Electromagnetics
CSIR-National Aerospace Laboratories
Bangalore, Karnataka
India

ISSN 2191-8112   ISSN 2191-8120   (electronic)
SpringerBriefs in Electrical and Computer Engineering
ISSN 2365-6239   ISSN 2365-6247   (electronic)
SpringerBriefs in Computational Electromagnetics
ISBN 978-981-287-801-4   ISBN 978-981-287-802-1   (eBook)
DOI 10.1007/978-981-287-802-1

Library of Congress Control Number: 2015948162

Springer Singapore Heidelberg New York Dordrecht London

Springer Science+Business Media Singapore Pte Ltd. is part of Springer Science+Business Media
(www.springer.com)

To Professor Vinod K. Tewary

In Memory of Dr. Rakesh Mohan Jha
Great scientist, mentor, and excellent
human being

Dr. Rakesh Mohan Jha was a brilliant contributor to science, a wonderful human being, and a great mentor and friend to all of us associated with this book. With a heavy heart we mourn his sudden and untimely demise and dedicate this book to his memory.

Foreword

National Aerospace Laboratories (NAL), a constituent of the Council of Scientific and Industrial Research (CSIR), is the only civilian aerospace R&D Institution in India. CSIR-NAL is a high-technology institution focusing on various disciplines in aerospace and has a mandate to develop aerospace technologies with strong science content, design and build small and medium-size civil aircraft prototypes, and support all national aerospace programs. It has many advanced test facilities including trisonic wind tunnels which are recognized as national facilities. The areas of expertise and competencies include computational fluid dynamics, experimental aerodynamics, electromagnetics, flight mechanics and control, turbo-machinery and combustion, composites for airframes, avionics, aerospace materials, structural design, analysis, and testing. CSIR-NAL is located in Bangalore, India, with the CSIR Headquarters being located in New Delhi.

CSIR-NAL and Springer have recently signed a cooperation agreement for the publication of selected works of authors from CSIR-NAL as Springer book volumes. Within these books, recent research in the different fields of aerospace that demonstrate CSIR-NAL's outstanding research competencies and capabilities to the global scientific community will be documented.

The first set of five books are from selected works carried out at the CSIR-NAL's Centre for Electromagnetics and are presented as part of the series SpringerBriefs in Computational Electromagnetics, which is a sub-series of SpringerBriefs in Electrical and Computer Engineering.

CSIR-NAL's Centre for Electromagnetics mainly addresses issues related to electromagnetic (EM) design and analysis carried out in the context of aerospace engineering in the presence of large airframe structures, which is vastly different and in contrast to classical electromagnetics and which often assumes a free-space ambience. The pioneering work done by the Centre for Electromagnetics in some of these niche areas has led to founding the basis of contemporary theories. For example, the geodesic constant method (GCM) proposed by the scientists of the Centre for Electromagnetics is immensely popular with the peers worldwide, and forms the basis for modern conformal antenna array theory.

The activities of the Centre for Electromagnetics consist of (i) Surface modeling and ray tracing, (ii) Airborne antenna analysis and siting (for aircraft, satellites and SLV), (iii) Radar cross section (RCS) studies of aerospace vehicles, including radar absorbing materials (RAM) and structures (RAS), RCS reduction and active RCS reduction, (iv) Phased antenna arrays, conformal arrays, and conformal adaptive array design, (v) Frequency-selective surface (FSS), (vi) Airborne and ground-based radomes, (vii) Metamaterials for aerospace applications including in the Terahertz (THz) domain, and (viii) EM characterization of materials.

It is hoped that this dissemination of information through these SpringerBriefs will encourage new research as well as forge new partnerships with academic and research organizations worldwide.

Shyam Chetty
Director
CSIR-National Aerospace Laboratories
Bangalore, India

Preface

The increasing demand of high data rates for wireless space communications has resulted in exploring of an unallocated frequency spectrum, viz., the terahertz spectrum, ranging from 300 GHz–30 THz. Space communications systems using THz spectrum can resolve the problems of limited bandwidth and spectrum scarcity of present wireless communications. Further, radio frequency interference can also be avoided by using THz frequencies for space applications. Apart from providing wider bandwidth, other advantages of THz spectrum for space communication are reduced-size antennas and low-power devices compared to that of microwave frequencies.

Although the use of THz frequencies for space applications has many advantages, the advancement in this region is slow due to the lack of suitable devices in THz range. The recent progress in semiconductor technology has resulted in development of terahertz devices, which has increased the interest of scientists in exploring this terahertz region for wireless space communications.

There are some factors that limit application of THz waves for communication. They are high atmospheric attenuation and high free-space path loss, which limit its application for short range communications. However, atmospheric attenuation of THz waves is not an issue for space applications as it is atmosphere-free. The high free-space path loss of THz waves can be compensated by using an antenna with high gain and directivity. Thus, the main aim of this brief is to design an antenna with high gain and high directivity for aerospace platform.

Microstrip patch antennas are low cost, light weight, portable, conformal, and can be easily installed and fabricated. These characteristics of microstrip patch antenna have proven to be advantageous for wireless space applications. However, they have a disadvantage of low bandwidth and low gain. The performance of the patch antenna can be improved by increasing the size of the patch and using thick substrate with low dielectric permittivity. The use of thicker substrate however results in surface wave loss because most of the radiated power of the patch antenna is trapped within the substrate. This loss can be overcome by the use of photonic band gap (PBG) substrate, which is obtained by implanting air gap cylinders periodically on a substrate. This reduces the effective dielectric permittivity and attenuates wave propagation in certain frequency band gaps. Thus the gain and directivity of the microstrip patch antenna can be improved by using a PBG substrate.

In this brief, design of such high-gain antennas and their performance enhancement using various mechanisms has been provided. Further, optimization of antenna models using multi-objective evolutionary algorithm has been described. The designed antenna may be used for various wireless applications, such as aircraft collision avoidance system, global positioning systems, telemetry, on-vehicle satellite links, missile radars, inter-orbital communications, communications inside space vehicle, etc., because of its compact size and enhanced performance characteristics.

Balamati Choudhury
Bhavani Danana
Rakesh Mohan Jha

Acknowledgments

We would like to thank Mr. Shyam Chetty, Director, CSIR-National Aerospace Laboratories, Bangalore for his permission and support to write this SpringerBrief.

We would also like to acknowledge valuable suggestions from our colleagues at the Centre for Electromagnetics, Dr. R.U. Nair, Dr. Hema Singh, Dr. Shiv Narayan, and Mr. K.S. Venu and their invaluable support during the course of writing this book.

But for the concerted support and encouragement from Springer, especially the efforts of Suvira Srivastav, Associate Director, and Swati Mehershi, Senior Editor, Applied Sciences & Engineering, it would not have been possible to bring out this book within such a short span of time. We very much appreciate the continued support by Ms. Kamiya Khatter and Ms. Aparajita Singh of Springer towards bringing out this brief.

Balamati Choudhury
Bhavani Danana
Rakesh Mohan Jha

Contents

About the Authors

Dr. Balamati Choudhury is currently working as a scientist at Centre for Electromagnetics of CSIR-National Aerospace Laboratories, Bangalore, India since April 2008. She obtained her M.Tech. (ECE) degree in 2006 and Ph.D. (Engg.) degree in Microwave Engineering from Biju Patnaik University of Technology (BPUT), Rourkela, Orissa, India in 2013. During the period of 2006–2008, she was a Senior Lecturer in Department of Electronics and Communication at NIST, Orissa India. Her active areas of research interests are in the domain of soft computing techniques in electromagnetics, computational electromagnetics for aerospace applications and metamaterial design applications. She was also the recipient of the CSIR-NAL Young Scientist Award for the year 2013–2014 for her contribution in the area of Computational Electromagnetics for Aerospace Applications. She has authored and co-authored over 100 scientific research papers and technical reports including a book and three book chapters. Dr. Balamati is also an Assistant Professor of AcSIR, New Delhi.

Ms. Bhavani Danana has obtained her M.Tech. degree from Jawaharlal Nehru Technical University, Kakinada. During the course of her studies, she interned as a visiting student at the Centre for Electromagnetics, CSIR-National Aerospace Laboratories (CSIR-NAL), and worked on terahertz antennas, photonic band gap structures, etc.

Dr. Rakesh Mohan Jha was Chief Scientist & Head, Centre for Electromagnetics, CSIR-National Aerospace Laboratories, Bangalore. Dr. Jha obtained a dual degree in BE (Hons.) EEE and M.Sc. (Hons.) Physics from BITS, Pilani (Raj.) India, in 1982. He obtained his Ph.D. (Engg.) degree from Department of Aerospace Engineering of Indian Institute of Science, Bangalore in 1989, in the area of computational electromagnetics for aerospace applications. Dr. Jha was a SERC (UK) Visiting Post-Doctoral Research Fellow at University of Oxford, Department of Engineering Science in 1991. He worked as an Alexander von Humboldt Fellow at the Institute for High-Frequency Techniques and Electronics of the University of Karlsruhe, Germany (1992–1993, 1997). He was awarded the Sir C.V. Raman Award for Aerospace Engineering for the Year 1999. Dr. Jha was elected Fellow of

INAE in 2010, for his contributions to the EM Applications to Aerospace Engineering. He was also the Fellow of IETE and Distinguished Fellow of ICCES. Dr. Jha has authored or co-authored several books, and more than five hundred scientific research papers and technical reports. He passed away during the production of this book of a cardiac arrest.

Abbreviations

DGS   Defected ground structure
NEP   Noise equivalent power
NET   Noise equivalent temperature
PBG   Photonic band gap
PC   Photonic crystal
QCL   Quantum cascade laser

List of Figures

List of Tables

PBG based Terahertz Antenna for Aerospace Applications

1 Introduction

The growing demand of high data rate wireless communication systems is bringing attention of researchers to the unallocated regime of the frequency spectrum i.e., the terahertz band. The terahertz band refers to the range of frequencies that lies between 0.1 and 10 THz, which also includes the submillimeter band. In contrast to its neighbors in the spectrum, growth in THz technology has been slow. However, this technology has been evolving rapidly to meet the great demand for high data rates in wireless communications. As per survey, the essential data rate may ascend to several tens of Gbit/s in the near future. To attain these high data rates, one feasible solution is to increase the available bandwidth, which is not possible below a frequency of 300 GHz, which is fully allocated. This demand is met by utilizing a new frequency spectrum i.e., THz spectrum, frequencies ranging from 300 GHz to 3 THz (Fig. 1), as it is unallocated (Fitch and Osiander 2004). Thus use of THz frequencies for communication can resolve the problems of spectrum scarceness and bandwidth limitations of present wireless systems. Further, the problem of interference in wireless communications can also be solved using THz frequencies.

The factor that limits the use of THz band for wireless communication is the high atmospheric attenuation. Despite the high atmospheric attenuation level, the terahertz finds its position in the space communication applications where atmospheric absorption is not an issue.

The main advantage of THz spectrum in space communication is that this frequency band provides greater communication bandwidth compared to microwave frequencies. Furthermore the important point is that this frequency spectrum is not in use unlike other bands which are fully allocated.

A few advantages of terahertz technology in space communication have been mentioned below for better understanding of the latest trends in this platform:

© The Author(s) 2016
B. Choudhury et al., *PBG based Terahertz Antenna for Aerospace Applications*,
SpringerBriefs in Computational Electromagnetics,
DOI 10.1007/978-981-287-802-1_1

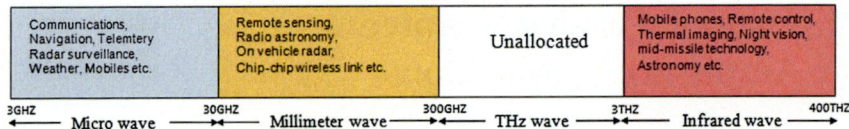

Fig. 1 Frequency spectrum

- The THz band has multiple GHz channel bandwidth, which provides high data rates with less power consumption and higher channel capacity.
- The system design at THz frequency requires small size antennas since the wavelength is in submillimeter range. The smaller transmitting and receiving antenna allows using smaller space vehicles. This results in reduction of size and weight of space vehicles which provide an added advantage for space applications.
- Terahertz frequencies are highly affected by atmospheric attenuation. But in space applications, due to atmospheric free environment, THz waves will not be affected by attenuation. Moreover, atmospheric attenuation for THz waves is less when compared to IR waves and laser signals.
- High imaging resolution is provided at THz frequencies than at microwave frequencies.
- Communications at THz frequencies are highly directional than at microwave frequencies as THz wave diffracts less than the microwave.
- The maximum transmission range of a 60 GHz system and comparatively less power 0.4 THz system is almost same and is equal to 2 km (approx.) (Federici and Moeller 2010). Thus systems at THz frequencies employ low power devices when compared to microwave frequencies.
- Secure communications are possible at THz frequencies as highly directional and less scattered communications in this range reduce the area of exposure for signal detection.
- Scintillation effect for THz wave is less when compared to microwave.
- Interference from ground-based transmitters and receivers can be avoided due to the atmospheric attenuation of THz signals. Because THz signals from an aircraft in space will be attenuated as they reach the ground, making it difficult to be detected by a ground-based receiver. Similarly, signals from ground-based transmitters, also suffer atmospheric attenuation as they reach space and cannot be detected by receivers in space. Thus the problem of terrestrial radio frequency interference can be avoided at THz frequencies.
- THz waves are nonionizing and hence are not harmful for human body.

All the above-mentioned advantages increased the scope of utilizing THz frequencies for space communications. Hence, in this work, the instruments required for THz space communication has been explored along with design optimization of THz antennas.

2 Challenges in Terahertz Technology

Although the applications of terahertz technology in radio astronomy, atmospheric earth observation science, and planetary/cometary science have been initiated in last decades, the THz band is not yet keeping pace because of lack of suitable devices and sources. At THz frequencies, skin depth is small, resulting in more conductive loss. Conventional antennas that already exist are not suitable for communications involving THz frequencies.

The free space path loss is given by $L_o = \left(\frac{4\pi d}{\lambda}\right)^2$, the above equation which shows that the free space path loss is inversely proportional to the square of the wavelength. At terahertz frequencies wavelengths are very small ranging from 3 mm to 30 µm which results in very high free space path loss. This limits its application to short-range communications. Space craft link analysis suggests that high-power transmitters are required for long-range communications (Hwu et al. 2013). Hence, the most important challenge is to design an antenna with high gain and high directivity to balance for high propagation losses at THz frequencies. Furthermore, the other challenges like fabrication issues, material issues are also of concern because of the natural breakpoint in material properties at 1–3 THz (Smith et al. 2004).

3 Trends in THz Space Communication Systems

Siegel (2010) termed THz for space applications as golden age as it offers unique spectral measurements in space. Thermal emissions from lightweight molecules and atoms, optical energy from distant galaxies fall in THz band requiring THz instruments. THz measurements provide data of our atmosphere and other planets, asteroids, moon, sun, etc., which help us in studying the dynamic processes taking place in our universe. Measurements of space-based THz instruments unlike ground-based instruments will not be disturbed by atmospheric absorption, attenuation, and turbulence. The ability to see through dust, clouds, etc., has provided an added advantage for utilizing THz waves in space applications. Success of Spitzer, Herschel, and many other THz missions have resulted for more future proposals for THz space applications. The author expects many similar proposals to be likely funded in the future as the interest in this frequency region is mounting.

Usage of THz region had already been in practice for astronomy applications in addition to spectral measurements. However, recent studies and developments in THz region increased the interest of scientists in exploring the usage of these frequencies for space communications like interorbital communication, space vehicle interior wireless communication, satellite communication, etc.

THz wireless communication inside space vehicles is gaining interest as it provides high data rates. The other advantage is that it requires small size antennas and eliminates the use of cables and wires. This reduces the weight and size of space vehicles and hence requires less fuel which helps in reducing the operational cost of the vehicle. The problem of radio frequency interference from ground-based equipment is also reduced due to the use of THz frequencies and thus the loss of signal can be reduced.

The demand for high data rates for deep space communication is also increasing and can be met using THz frequency band. The gain of an antenna is directly proportional to the square of the carrier frequency used. Hence the use of THz carrier frequencies provide high gain antennas when compared to RF frequencies. And also, THz communication in space is not affected by attenuation and interference which results in reliable communication without signal degradation. Power received by a receiver antenna is given by Eq. (1)

$$P_\mathrm{r} = \frac{P_\mathrm{t} A_\mathrm{t} A_\mathrm{r}}{r^2 \lambda^2} \tag{1}$$

where
P_r power received
P_t power transmitted
A_t effective area of transmitter
A_r effective area of receiver

Hence it is observed that communications using THz spectrum require low power transmitters due to its smaller wavelengths.

Hwu et al. (2013) explored the potentials applications of terahertz wireless communication for space craft interior (Fig. 2) as well as for interorbital data transfer (Fig. 3). It has been reported that 10 Gbps per GHz data rate is achievable at THz frequencies.

Han et al. (2015) proposed a THz space application system to be used in intersatellite links which helps in reducing the free space path loss in atmosphere. It was reported that a data rate of 10 Gbps can be achieved using this design. The application challenges in terahertz antenna have also been demonstrated.

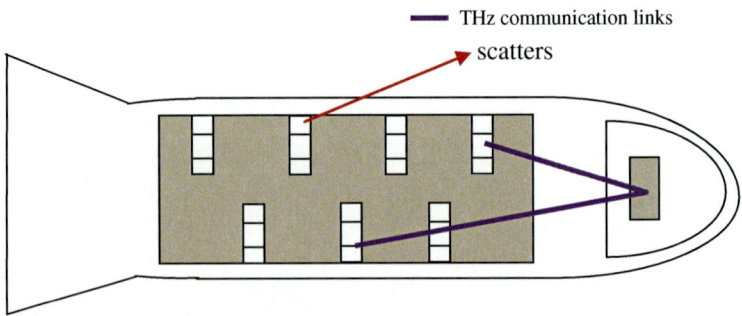

Fig. 2 THz communication inside space vehicle

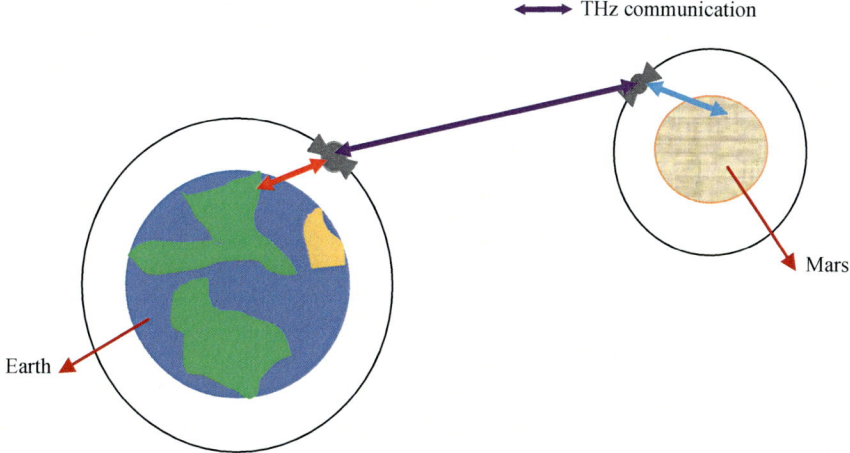

Fig. 3 Interorbital communication

4 Developments in THz Devices

Although the terahertz spectrum has tremendous potential in interorbital space communications and indoor communications including security applications, the development is slow because of lack of terahertz sources and detectors. But the recent progress in semiconductor technologies resulted in development of THz devices. In these sections, a brief review of development of THz devices has been reported.

4.1 THz Sources

In earlier days, THz waves were generated from thermal sources using Fourier transform infrared spectrometers (Fitch and Osiander 2004). But from 1975, laser sources were used to generate THz waves. Molecular gas laser can be used for generating THz wave, but these systems are expensive. THz wave generation using semiconductor lasers is inexpensive but requires cryo-cooling. Lead-salt lasers can also generate THz wave, but only at low temperatures and in the presence of magnetic field. All these disadvantages can be overcome using femtosecond laser.

When a femtosecond laser pulse (Zhang 2004) is focused on a photoconductive region of a photoconductive dipole antenna, charge carriers are produced which are accelerated due to bias voltage applied to the antenna resulting in generation of a time-varying current pulse (Fig. 4). This time-varying current pulse results in generation of a THz wave according to Maxwell's principle.

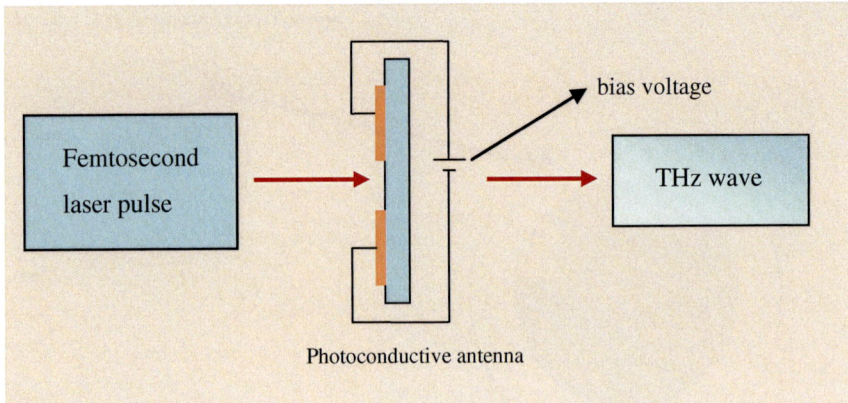

Fig. 4 THz wave generation

Frequency domain or continuous wave THz systems use photomixing process (Fig. 5) to generate a THz wave at a frequency equal to the frequency difference of two continuous wave laser beam sources that are mixed. But the use of this process for generating frequencies above 1 THz is not so efficient (Fitch and Osiander 2004). This process can be used to generate THz wave in the frequency range of 0.3–0.6 THz.

In recent developments, the two laser sources were replaced by a single laser beam source hence making this process more compact and cheap.

Backward wave oscillator can generate THz signals below 1 THz. High-power sources are required for long-range communications. Use of quantum cascade laser (QCL) can generate an output power of 90 mW, but only above 1 THz. Federici and Moeller (2010) suggested an alternate process to generate THz wave using

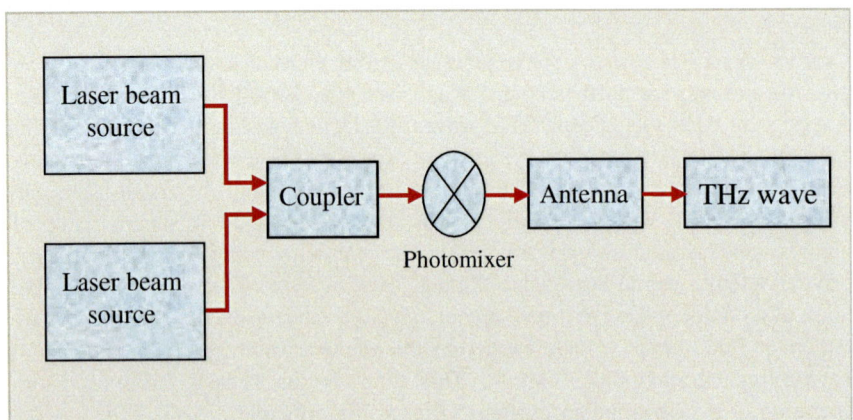

Fig. 5 Photomixing process

microwave frequency multipliers and solid-state devices where frequency multiplied microwave sources are used to generate the wave in THz frequency range.

4.2 THz Detectors and Receivers

The detection techniques of THz signals are generally based on heterodyne principle (generating a new frequency by combining two or more signals of different frequencies in devices such as a vacuum tube, diode mixer, and transistor.). In many cases, THz detection techniques are implemented by reversing the THz generation techniques.

Hagmann et al. (1992) designed a bolometric receiver for space applications in THz range that works in a band of wavelength extending from 200 to 700 µm and reported that a noise equivalent power (NEP) of 5×10^{-17} W/$\sqrt{\text{Hz}}$ can be achieved using this design at a temperature of 100 mK. The noise equivalent temperature (NET) of this receiver at 1 THz in a 10 % bandwidth is 1.8×10^{-5} K/$\sqrt{\text{Hz}}$, whereas it is 1.5×10^{-3} K/$\sqrt{\text{Hz}}$ for a heterodyne receiver at a frequency of 1 GHz, which proves that at these conditions the designed bolometric receiver is 100 times more sensitive than a normal heterodyne receiver.

Zimmermann et al. (1995) presented a design of heterodyne receiver that operates at a frequency of 1 THz. This design uses waveguide mixer and schottky diode frequency multipliers. The output power of the local oscillator achieved was 70 mW ± 5 %. Cherednichenko et al. (2002) reported a low noise 1.6 THz heterodyne receiver using hot electron bolometer mixer with phonon cooling system. This mixer is designed using a superconducting NbN of thickness 3.5 nm. In 1–2 GHz band, the receiver performance was tested using twin-slot antenna and spiral antenna at 2 K and reported that at 1.6 THz the receiver noise temperature was 700 K. Similarly in 3–8 GHz band at 4.2 K the receiver noise temperature was measured as 1500 K using spiral antenna. Later, Karpov et al. (2003) discussed the advancement in the design of SIS mixer of the heterodyne spectrometer at a band of 1.1–1.25 THz and reported that a receiver noise of 500 K can be achieved with this design. This SIS mixer design is made using two junctions of Nb/AlN/NbTiN. Maiwald et al. (2004) presented a design for solid-state THz frequency multiplier chains for space application receivers.

Schottky diodes have been emerged as highly sensitive detectors in THz range when incoherent terahertz sources are used. Sobis et al. (2013) reported a model of heterodyne receiver using GaAs schottky diode with monolithically integrated technology for a frequency range of 300–1.2 GHz. Chouvaev et al. (1998) presented a model of normal metal hot electron micro-bolometer that could achieve an NEP of 5×10^{-18} W/$\sqrt{\text{Hz}}$ at a temperature of 300mK. This design uses Andreev mirrors and SIN junctions for electronic cooling. Later Karasik et al. (2005) discussed the theory of a hot electron bolometer that is made of superconducting Ti bridge and Nb Andreev contacts. Using this design, they could achieve an NEP of 10^{-20} W/$\sqrt{\text{Hz}}$ at 300 mK, whereas a micro-machined bolometer requires space instrument temperatures below 10 mK for achieving this sensitivity, which is difficult. Thus this design can reduce the difficulty of cryo-cooling in space application.

4.3   THz Antennas

Free space path loss is very high at terahertz frequencies. To improve the link budget of the system, antennas with high gain and high directivity are required. Highly directive antenna can overcome the propagation losses in the atmosphere and can transmit data over long distances. If the thickness of the substrate used in antennas is greater than the wavelength of operation, the radiated power gets trapped in the substrate. Hence a substrate with thickness much lesser than the wavelength is required, which is difficult to obtain and remains as a challenge in designing an antenna for THz frequencies.

THz radiations can be generated and detected using photoconductive antennas. Li and Hunag (2006) compared two different photoconductive antennas and suggested that better performance can be obtained using a sharp edged dipole antenna known as indentation antenna than a normal dipole antenna. The electric filed generated in dipole antenna (Fig. 6) is identical to the electric field between two line charges, whereas in indentation antenna (Fig. 7), it is identical to the electrical field between two point charges. Hence electric field of indentation antenna is stronger than that of dipole antenna. Li and Hunag (2006) proved that the current is proportional to the square root of electric field using the results of two experiments conducted by Cai et al. (1997) and Zhang et al. (2004). The former reported that the THz wave radiated from indentation antenna is larger than that radiated from dipole antenna as current generated in indentation antenna is more than the current generated in dipole antenna.

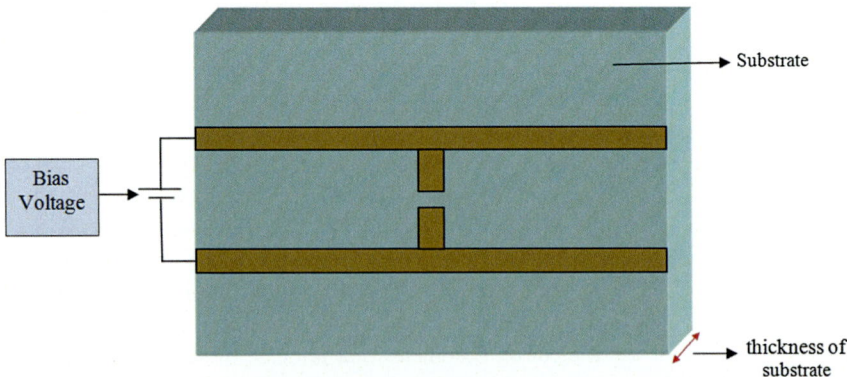

Fig. 6 Photo conductive dipole antenna

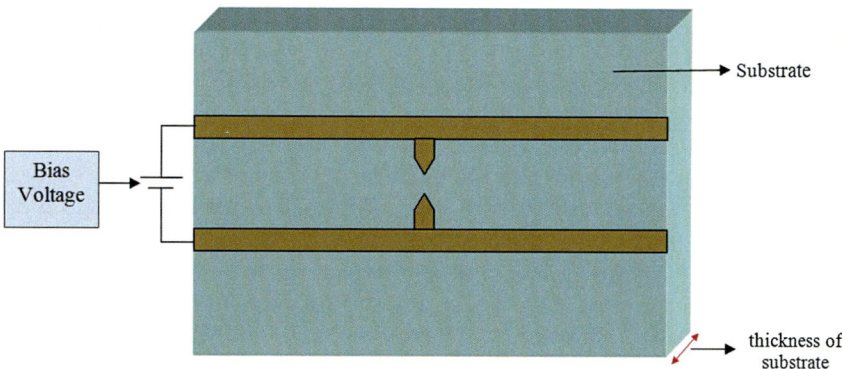

Fig. 7 Indentation antenna

Ezdi et al. (2009) , developed a hybrid model for pulse excited THz antennas and analyzed it in time domain to predict its behavior. The dipole antenna is designed using four Smith's traveling wave element.

A traveling wave element (Fig. 8) consists of source signal at one end which places a current pulse on the element resulting in radiation due to acceleration of charge. The pulse travels the entire length and gets absorbed by terminator which acts as a perfect absorber. The radiating system model consists of a metallic structure comprising contact pads, strip line, and dipole structure on a semiconductor substrate. When an ultrafast laser excitation is given to the semiconductor region between two arms, charge carriers are produced which are accelerated by the applied electric field through contact pads, resulting in the generation of short current pulse and thus generating THz radiation. Ezdi et al. (2009) measured the photo current by incorporating smith model with the model developed by Jepsen et al. (1996) and later by Piao et al. (2000). Thus an expression for electric radiation was derived which is dependent on original current pulse and the pulse reflected to the center of dipole from which terahertz wave generated can be determined.

Llombart et al. (2010) proposed a design of a reflector antenna system for high resolution THz imaging using confocal gregorian geometry, a rotating mirror, and a

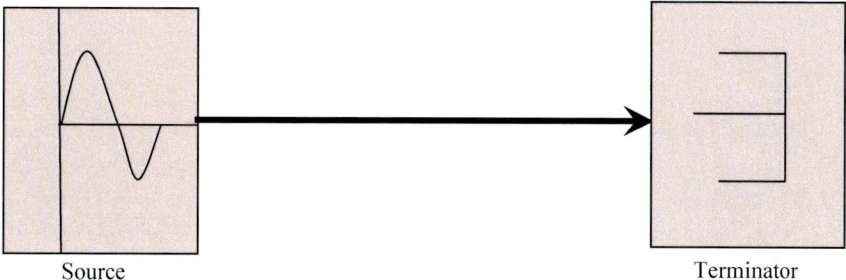

Source   Terminator

Fig. 8 Travelling wave element

feed. Here ellipsoid reflector surface is used instead of paraboloid reflector surface, for targets which are close to antenna aperture. The scanning speed of imagery is improved here using a small rotating mirror (phased array is used for space applications) to steer the beam while main reflector remains scanning the beam projected onto the target. Dimensions of antenna depends on magnification (M: ratio of focal distance of main and sub reflector) and f-number (Fm- ratio of focal distance and diameter of main reflector). More M results in more path length error, less M results in more size of rotating mirror, more Fm implies quite large system, and less Fm increases path length error. For refocusing, feed reflector is displaced instead of the entire secondary reflector system to reduce large beam distortions. Llombart et al. (2010) optimized the values of Fm (=1.2) and M (=10) and reported that this design scans 0.5 m beam at 25 m standoff range with <1 % increase in beam width. The results were analyzed using GRASP simulation.

Singh et al. (2009) analyzed the properties of spiral-type terahertz antenna and reported resonant eigen modes for different number of spiral windings. Using Mushiake principle (property of self complementary antenna), they analyzed that at higher frequencies the incident intensity is equally divided into reflected and transmitted wave.

Pitra et al. (2013) designed a circularly polarized THz antenna. To radiate high power, LTGaAS photomixer with interdigital electrode (IDE) is integrated on the antenna and good impedance matching is obtained by designing a matching circuit with precisely calculated capacitance of IDE. By varying the capacitance of IDE, the system can be tuned to operate at a given frequency. The dimensions of required square patch antenna were reported. This antenna is placed on a silicon substrate lens and excited by means of cross-slot aperture. The numerical model of this antenna is simulated in transient solver of FEM-based software with a frequency sweep ranging from 0.4 to 1.6 THz in time domain with a discrete port excitation source. Pitra et al. (2013) reported that this design could achieve a 3 dB cross polarization bandwidth of 20 % with 1 THz central frequency.

High Directivity and gain can be achieved using phased array antenna in which the phase excitation between the elements of array is varied using phase shifter to steer the direction of major radiation onto the target automatically with high precision. Hwu et al. (2013) proposed that a nanophotonic phased array consisting of 64×64 nanoantennas balanced in power and integrated on a silicon chip and can be used in nonpoint-to-point fixed wireless THz communications.

5 Terahertz Antenna Design

The aim of the project is to design an antenna for wireless communications in THz range that can be used in aerospace applications. The desired requirements of antenna are high gain, high directivity, high efficiency, small size, wide bandwidth, and low cost.

Photoconductive antennas can be used for such applications because of their advantages such as compact size and high tunability at room temperature. However they have a disadvantage of low output power due to poor impedance matching when connected to a photomixer.

This disadvantage can be overcome using microstrip patch antenna. The study of microstrip patch antennas is emerging in wireless communications due to its several advantages. The advantages of microstrip patch antenna and its design procedure are discussed in the next sections.

5.1 Microstrip Patch Antenna

Microstrip patch antenna has several advantages such as low cost, low weight, ease of fabrication and integration, compactness, and portability. Another important feature of microstrip antenna is that it can be conformed to any curved surface i.e., the body of aerospace structures. The other advantage of microstrip antennas is that they can be easily installed on aircraft structures and can be used for high-frequency applications. Proper impedance matching can be obtained by optimizing the feed position which results in high output power. The use of these antennas proved to be advantageous for aerospace applications where low weight, compact size, and easily installed antennas are requisite.

Hence in this work, a microstrip patch antenna is designed and optimized to obtain high gain, high radiation efficiency, and high directivity with low return loss in terahertz range that can be used for space applications.

5.1.1 Design Procedure

Design and simulation has been done using FEM-based software by calling it directly from MATLAB using VBA language.

- VBA language has been accessed from MATLAB using "invoke" command.
- One new project module was created in the FEM-based software.
- The template for the project was selected as "Antenna (Planar)."
- By selecting 'Solve' → 'Units', units of frequency and dimensions were set to "THz" and "μm," respectively.
- The materials required for the design were loaded into Materials in FEM-based software navigation tree by right clicking on 'Materials' icon and selecting 'Load from material library.' If the required materials were not found, then the material with required dielectric constant can be created by right clicking on 'Materials' and selecting 'New material.'
- The Frequency range for which the simulation has to be done was specified by selecting 'Frequency range' icon.
- The FEM-based software file was saved with some valid name and closed.
- Now the code for designing and simulation was written in MATLAB editor window.

- Now the Matlab program was simulated which opens the existing FEM-based software file, designs the model, and performs the simulation.
- Using this FEM-based software, the radiation pattern, gain, directivity, return loss, and all other performance characteristics of a designed antenna model can be obtained.

Thus the design of antenna was done by interfacing FEM-based software and MATLAB and the simulated results were obtained.

5.1.2 Design of THz Patch Antenna

Microstrip patch antenna (Fig. 9) consists of a ground plane residing below the substrate. The patch and microstrip feed line are etched above the substrate. The radiation in microstrip patch antenna is mainly due to fringing fields that arise between patch and ground plane.

The ground plane should be a perfect electric conductor and hence PEC material is used in the design. The patch and the feed line are made of highly conducting material such as copper. The patch of this design is made in rectangular shape as it is easy to analyze and involve simple calculations.

The dimensions of microstrip patch antenna can be determined using the design equations through Eqs. (2)–(10) as given below.

Width of the patch

$$W_p = \frac{c}{2f_r\sqrt{\frac{\varepsilon_r+1}{2}}} \tag{2}$$

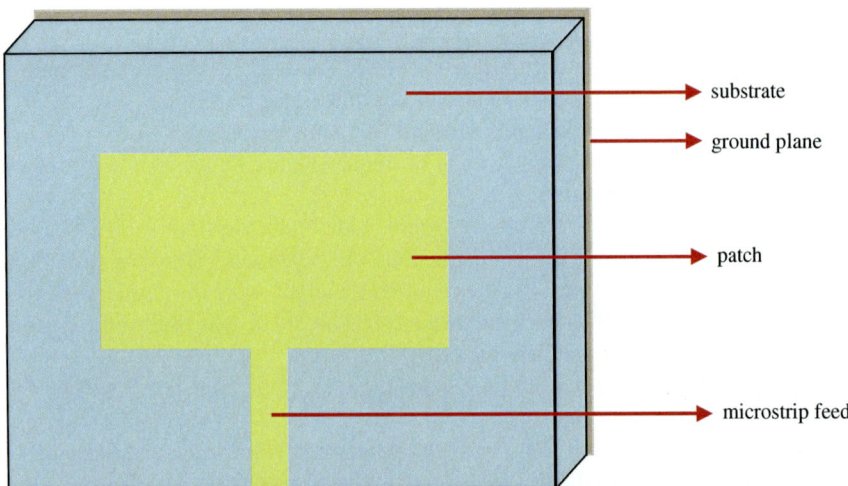

Fig. 9 Rectangular microstrip patch antenna

Width of the patch affects the bandwidth of the antenna. An increase in patch width results in an increase in the bandwidth of the antenna. However, larger patch width results in grating lobes.

Effective dielectric constant

$$\varepsilon_{reff} = \frac{\varepsilon_r + 1}{2} + \frac{\varepsilon_r - 1}{2}\left(1 + \frac{12 \times h}{W_p}\right) \tag{3}$$

Effective dielectric constant is calculated in order to include the effect of fringing fields which act outside the radiating patch. Due to these fringing fields, the patch appears to be longer than its actual length. Hence the actual length of the patch is considered by including the effect of these fringing fields.

Length of the patch

$$L_p = L_{eff} - (2 \times \Delta L) \tag{4}$$

where

$$L_{eff} = \frac{c}{2f_r\sqrt{\varepsilon_{reff}}} \tag{5}$$

$$\Delta L = 0.412h\left(\frac{\varepsilon_{reff} + 0.3}{\varepsilon_{reff} - 0.258}\right)\left(\frac{\frac{W_p}{h} + 0.264}{\frac{W_p}{h} + 0.8}\right) \tag{6}$$

Length of ground plane

$$L_g = L_p + (6 \times h) \tag{7}$$

Width of ground plane

$$W_g = W_p + (6 \times h) \tag{8}$$

The width of the microstrip feed line was calculated so as to supply a characteristic impedance of 50 Ω.

Width of microstrip feed line

$$W_f = h\left(\frac{377}{50\sqrt{\varepsilon_r}} - 2\right) \tag{9}$$

where
c velocity of light $= 3 \times 10^8$ m/s
f_r resonant frequency
ε_r dielectric constant of the substrate
h height of the substrate

Fig. 10 Microstrip patch
antenna with inset feed

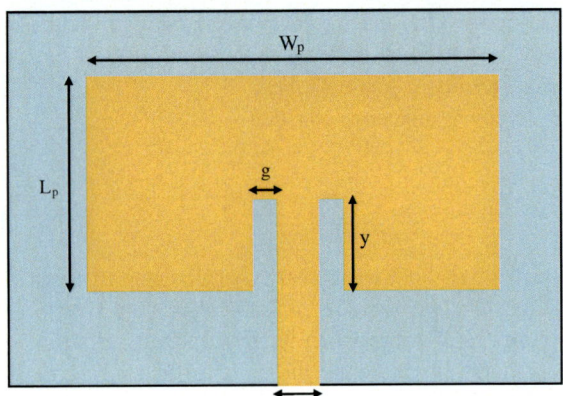

When the microstrip feed line is fed at the edge of the patch, the input impedance offered by the patch is very high typically ranging from 150 to 300 Ω when compared to the characteristic impedance of feed line which is of 50 Ω. This results in poor impedance matching which causes more reflections and loss of signal due to which output power decreases. This problem can be solved using inset feed technique in which the feed is inserted into the patch by creating a gap as shown in Fig. 10.

The input impedance of patch when the feed is inserted at a distance of 'y' from the edge is given by

$$Z_y = Z_{\text{edge}} \times \cos^2\left(\frac{\prod y}{L_p}\right) \tag{10}$$

where
Z_y input impedance of patch at distance 'y' form edge
Z_{edge} input impedance at the edge of the patch
L_p length of the patch

From the above formula, it is observed that as the feed is moved toward the center of the patch, the input impedance of the patch decreases, and at the center of the patch, the input impedance becomes zero. Hence the feed line inset position is to be adjusted in between the center of the patch and edge of the patch so as to obtain good impedance matching between patch and the feed line of characteristic impedance 50 Ω.

Kumar et al. (2013) demonstrated the effect of three different feeding techniques on radiation characteristics of microstrip patch antenna and analyzed the results. It was reported that antenna with cut feed technique provides efficient results when compared to antenna with other feeding techniques. The gain of antenna with inset cut feed is higher.

Table 1 Dimensions of rectangular microstrip patch antenna

Design parameters	Dimensional value (µm)
Design parameters	Dimensional value (µm)
Width of the patch (W_p)	151.5
Length of the patch (L_p)	119.54
Width of the ground and substrate	211.5
Length of the ground and substrate	179.54
Width of the feed (W_f)	23.53
Inset feed position (y)	38.64
Notch gap (g)	10

The required resonating frequency of the design is maintained at 0.7 THz. The dielectric constant and the thickness of the substrate are considered as 3 and 10 µm respectively. The thickness of patch and microstrip feed is 5 µm.

Hence the dimensions of the microstrip patch antenna are calculated using the design equations Eqs. (2)–(10) as given above and the obtained values are tabulated in Table 1. The rectangular microstrip patch antenna is designed using planar technology as it offers higher integration possibility (Jha and Singh 2010). The microstrip patch antenna model is designed using the dimensions mentioned in Table 1 and simulated using FEM-based software in-conjunction with MATLAB interfacing.

The performance characteristics of the designed microstrip patch antenna can be observed from Figs. 11, 12, 13, 14 and 15. The antenna resonates at a frequency of

Fig. 11 Return loss characteristics of microstrip patch antenna

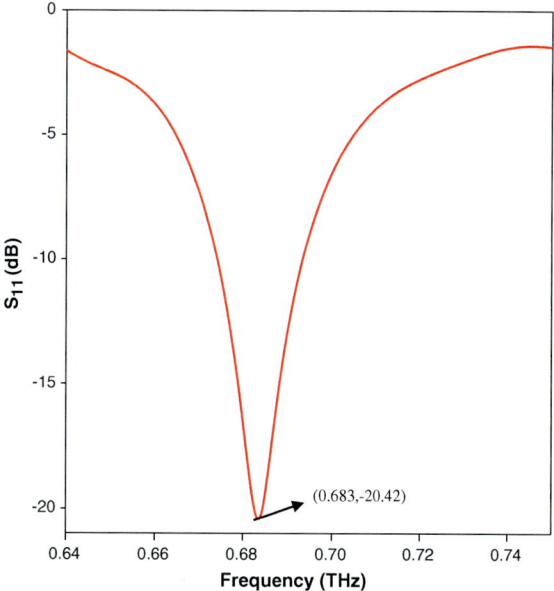

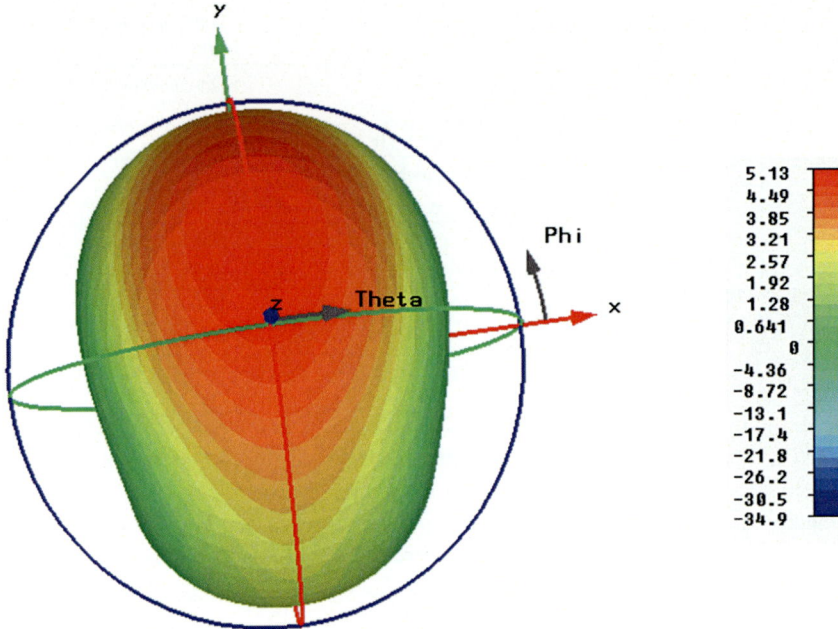

Fig. 12 3-D radiation pattern (gain) of microstrip patch antenna

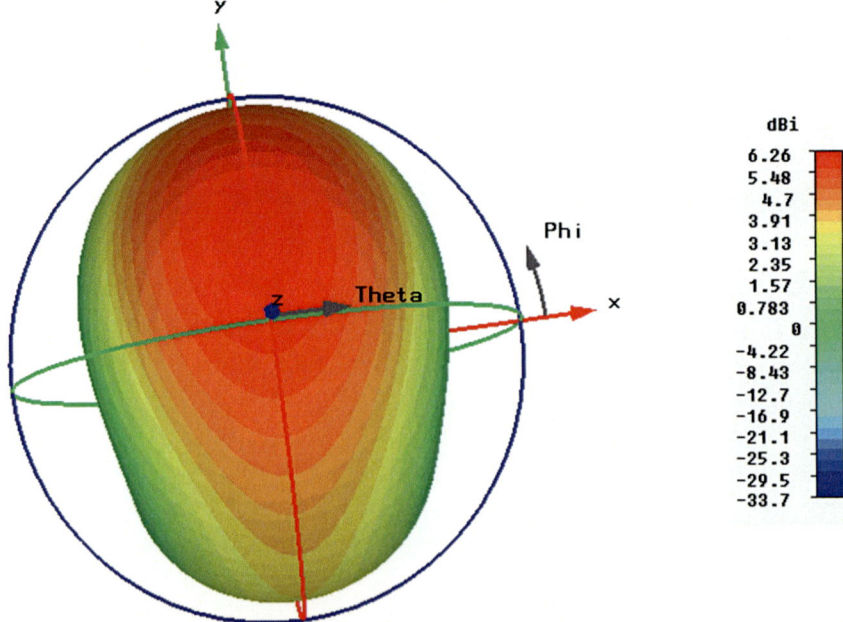

Fig. 13 3-D radiation pattern (directivity) of microstrip patch antenna

Fig. 14 Polar plot (phi = 90) of microstrip patch antenna

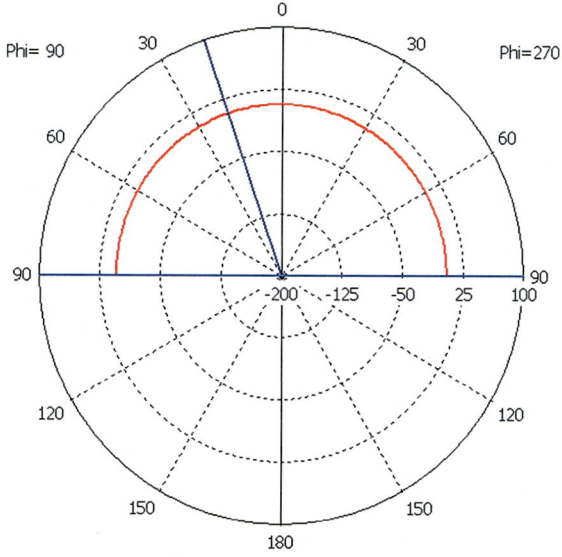

Theta / Degree vs. dBi

Fig. 15 Polar plot (phi = 0) of microstrip patch antenna

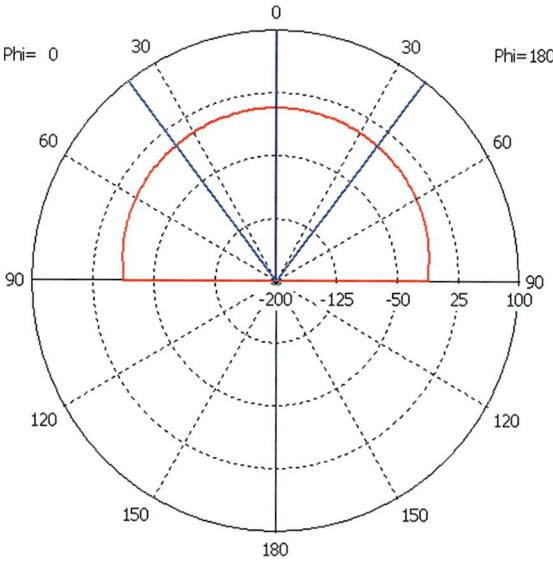

Theta / Degree vs. dBi

Table 2 Performance parameters of rectangular microstrip patch antenna	Frequency	0.683 THz
	Return loss	−20.42 dB
	Gain	5.131 dB
	Directivity	6.263 dBi

0.683 THz with a return loss of −20.42 dB (Table 2). The gain, directivity, and bandwidth of the designed model are 5.131 dB, 6.263 dBi and 2.79 %, respectively.

From the above results, it can be observed that the designed antenna has low gain, low directivity, and narrow bandwidth. Low gain and low directivity make this antenna unsuitable for space applications. Also narrow bandwidth results in reduced channel capacity and reduced power handling capacity. A possible solution for increasing the performance of microstrip patch antenna is to increase the size of the patch. Hence the dimensions of the patch were optimized to improve the performance of patch antenna.

5.1.3 Optimization of Patch Antenna

The optimized dimensions of the rectangular microstrip patch antenna are shown in Figs. 16 and 17 which are given below.

The dimensions of the substrate used in this design are $800 \times 800 \times 200 \ \mu m^3$ and made of a material with a dielectric constant 9.1. The dimensions of rectangular patch are $588 \times 226 \ \mu m^2$. The patch is fed by a microstrip line of width 99.89 μm to provide a characteristic impedance of 50 Ω. The thickness of patch and microstrip feed is 36 μm. The feed is inserted at a distance of 38.64 μm in order to obtain an impedance matching of 50 Ω.

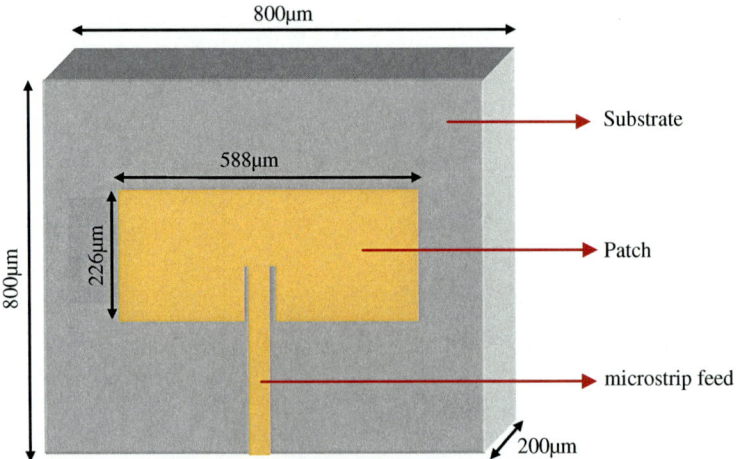

Fig. 16 Dimensions of optimized rectangular microstrip patch antenna

Fig. 17 Dimensions of
optimized microstrip feed

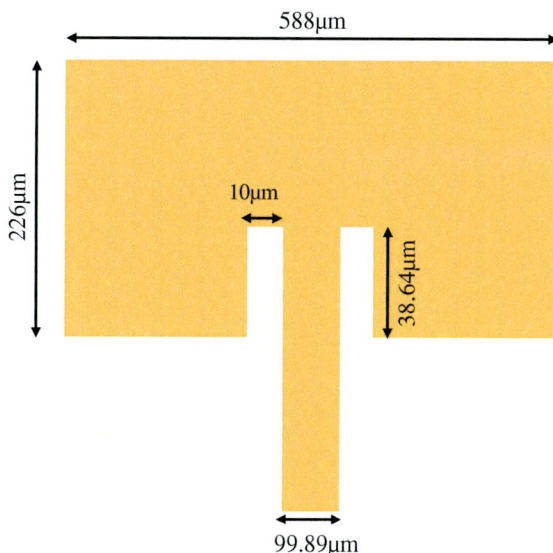

Table 3 Performance
parameters of optimized patch
antenna

Frequency	0.7066 THz
Return loss	−30.47 dB
Gain	9.759 dB
Directivity	10.28 dBi

The proposed model is designed and simulated in FEM-based software using MATLAB interfacing. The obtained results are given below.

From Table 3, it is observed that the designed antenna gives a return loss of −30.47 dB at a resonating frequency of 0.706 THz. The gain and directivity obtained at 0.7066 THz are 9.759 dB and 10.28 dBi, respectively.

From Figs. 18, 19, 20, 21 and 22, the performance of optimized patch antenna can be analyzed.

It is observed that the return loss, gain, and directivity of the designed microstrip patch were improved compared to the previous design. However, the obtained characteristics of antenna are very low for space applications. The radiation efficiency of this patch antenna is very less because most of the radiated power from conventional patch antenna is trapped in the dielectric substrate resulting in surface waves causing high power loss. This surface wave loss due to trapping increases with increase in thickness and dielectric permittivity of substrate material. About 60 % of the radiated power loss can occur when a thick substrate with high dielectric constant is used. Hence it is necessary to implement a feasible solution to improve the performance of microstrip patch antenna.

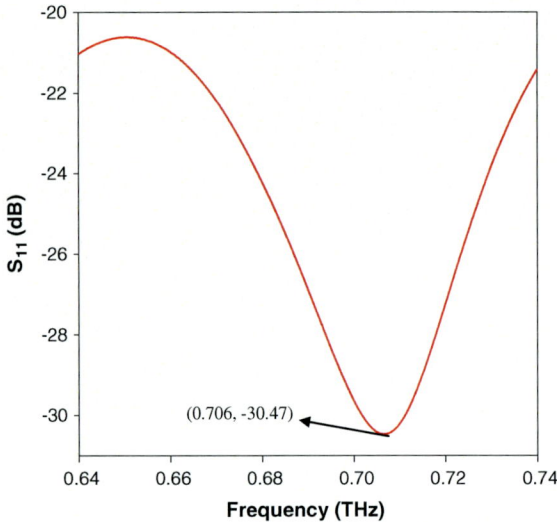

Fig.18 Return loss characteristics of optimized patch antenna

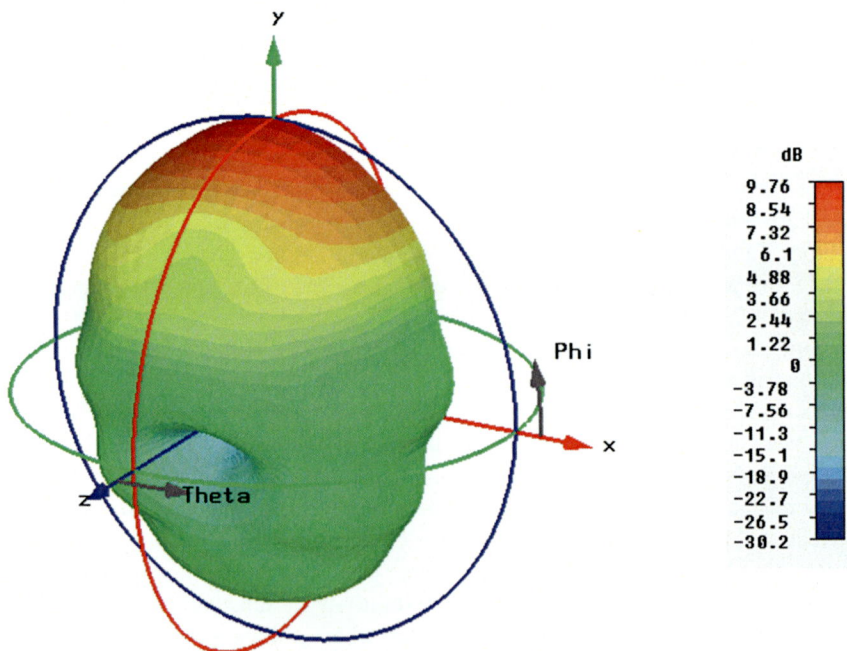

Fig. 19 3-D radiation pattern (gain) of optimized patch antenna

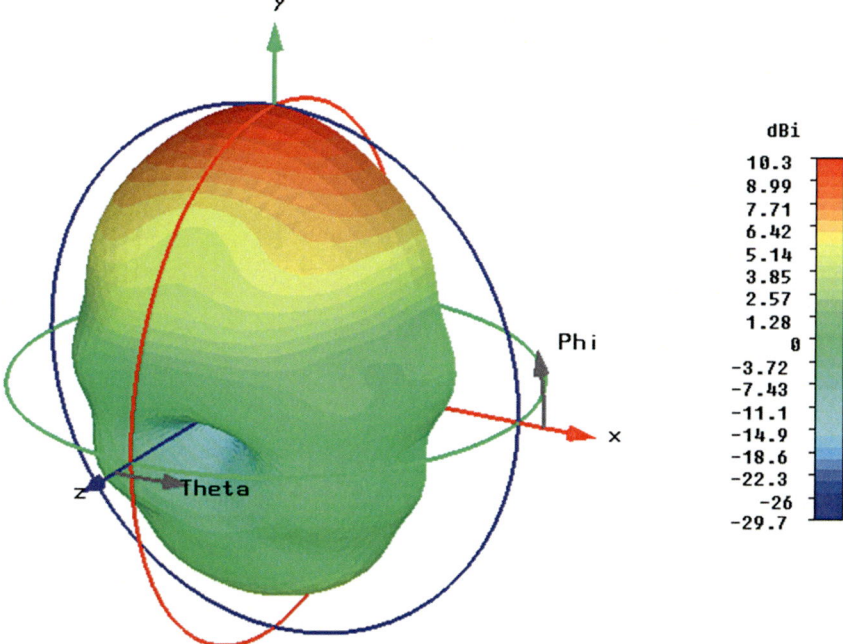

Fig. 20 3-D radiation pattern (directivity) of optimized patch antenna

5.2 *Performance Enhancement*

One possible solution is to use a thin substrate material with a low dielectric permittivity which reduces surface wave trapping. But this results in poor directivity (Jha and Singh 2010). Another possible solution to improve the gain is to use microstrip antenna array. But this results in bulky and heavier communication system which poses a constraint for space applications.

A lot of research work is going on to enhance the performance of patch antenna. A few promising solutions are mentioned in the next section.

5.2.1 Trends in Performance Enhancement of Patch Antenna

The performance of patch antenna can be improved using a defected ground structure and photonic band gap substrate. These methods help in enhancing the performance of microstrip patch antenna without implementing an added circuit that results in added weight and size. Hence using these methods better performance can be obtained from compact size microstrip patch antennas.

Fig. 21 Polar plot (phi = 90)
of optimized patch antenna

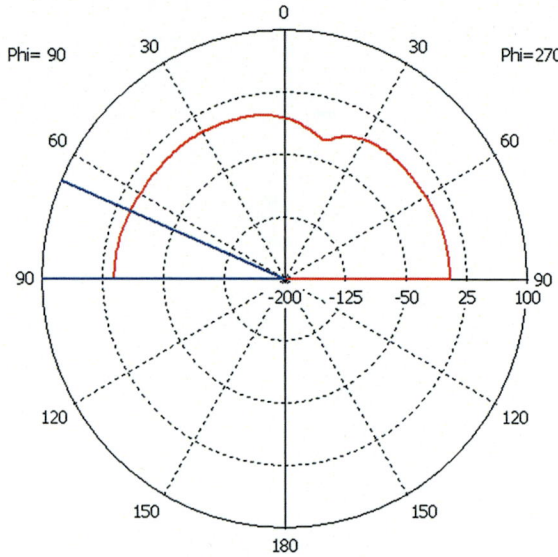

Theta / Degree vs. dBi

Fig. 22 Polar plot (phi = 0)
of optimized patch antenna

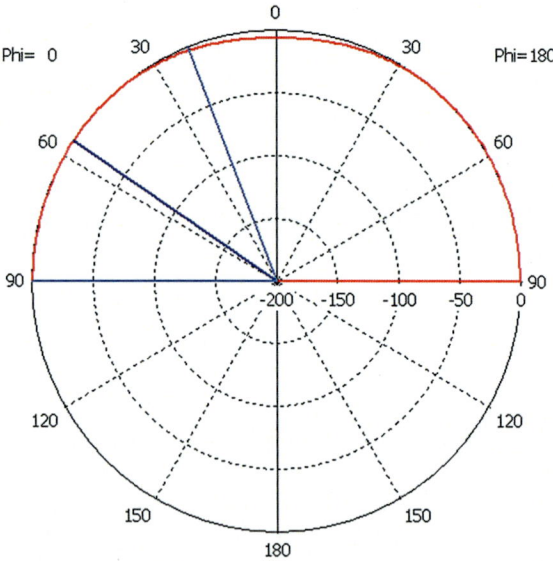

Theta / Degree vs. dBi

Defected Ground Structure (DGS)

Defected ground structures are used in microstrip patch antennas to improve return loss, bandwidth, and radiation efficiency. Some different-shaped slots are imprinted on the ground plane under the feed as shown in Fig. 23, which acts as LC resonator and affects the flow of antenna current. Thus the wave propagation through the substrate can be controlled and attenuated. By varying the shape, size, and position of the slot on the ground plane, the performance characteristics of microstrip patch antenna can be varied.

Babanrao et al. (2014) designed a microstrip patch antenna using line-shaped defected ground structure and reported that a return loss of −40.30 dB, and a radiation efficiency of 99 % could be achieved using this design. The designed antenna with DGS showed 73.70 % improvement in return loss and 6.36 % improvement in radiation efficiency over a microstrip patch antenna without DGS. However, the design had not shown any improvement in directivity.

Multilayer Substrate

In this type of antennas, the patch is etched on multilayer dielectric substrates where each layer of substrate is made of different material and of different thickness. Appropriate selection of material and thickness of substrate layers may result in reduction of surface wave loss thus improving the performance of antenna. Surface wave loss can be eliminated by placing a substrate layer with low-relative dielectric permittivity below the substrate layer with high-relative permittivity.

Sharma et al. (2009) designed a microstrip patch antenna on a multilayered substrate that can be used in THz frequencies. The designed antenna consists of a thicker middle layer and thin upper layer substrate, which helps in enhancing the bandwidth. It was reported that the designed antenna can provide a 10 dB impedance bandwidth of 33.67 %, a radiation efficiency of 90.69 %, and a gain of 10.05 dB.

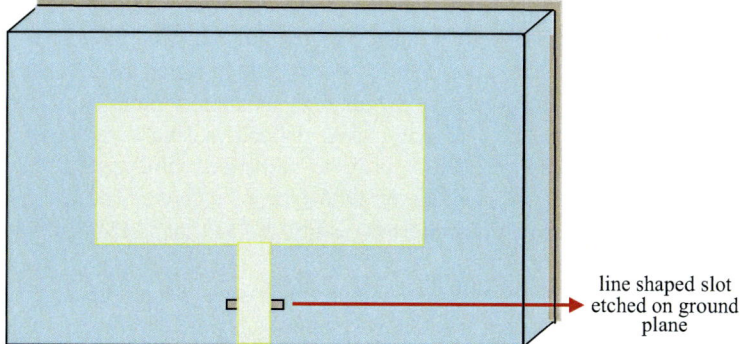

line shaped slot etched on ground plane

Fig. 23 Microstrip patch antenna with defected ground structure

Azarbar et al. (2014) designed a rectangular microstrip patch array antenna using multilayer technique. The designed antenna consists of a two layer substrate. The homogenous substrate with a dielectric permittivity of 2.08 is combined with a host material substrate with a dielectric permittivity of 3.82. Least square genetic algorithm is used to optimize the dimensions of the substrate. Multilayer substrate technique is used to enhance the bandwidth of the designed antenna. Patch array is used to enhance the antenna gain and directivity.

Puck Antenna

Puck antenna consists of a small dielectric superstrate i.e., a partially reflecting surface also known as puck which is placed in front of the feeding antenna. This puck allows the excited fields to reach the edges prior to leaking that enhances the antenna gain and bandwidth.

M. A. Al-Tarifi et al. (2015) designed a compact high gain antenna that operates in wide band width. It was reported that performance of antenna with circular shape puck is better when compared to the antenna with square shape puck, and its performance is four times more than traditional antennas with large superstrates. The size and shape of the ground are further optimized to obtain a small size antenna with better performance. It was reported that the designed puck antenna has an aperture efficiency of 51 %.

Photonic Band Gap (PBG) Substrate

A PBG substrate is obtained by implanting air gap cylinders periodically on a substrate (Fig. 24) due to which the effective dielectric permittivity of the substrate is reduced (Jha and Singh 2010). The bandwidth of a microstrip patch antenna using PBG substrate can be above 35 % of a conventional microstrip patch antenna (Tyagi and Vyas 2013). PBG structures attenuate wave propagation in a certain frequency band gaps. Due to these frequency band gaps, surface waves can be eliminated thus improving the bandwidth and directivity. PBG substrate can also be used to achieve negative effective dielectric permittivity. But microstrip patch antenna with negative dielectric permittivity results in high signal loss. Hence, it has been suggested that the permittivity of the antenna substrate should be positive and very less as compared to the frequently used dielectrics (Jha and Singh 2010). When an antenna is placed on a PBG substrate, radiations will be reflected in all directions resulting in improved radiation efficiency of antenna. The use of PBG substrate is favorable when a thick substrate is utilized to expand the bandwidth of the antenna.

Sharma et al. (2008) designed a rectangular microstrip patch antenna for THZ frequency (0.64–0.8 THZ) using photonic crystal as substrate which gives a gain of 3.852 dB and bandwidth of 13.36 % (FEM-based software simulation results). A strongly excited leaky wave gives high gain which is obtained when air blocks on

Fig. 24 PBG substrate

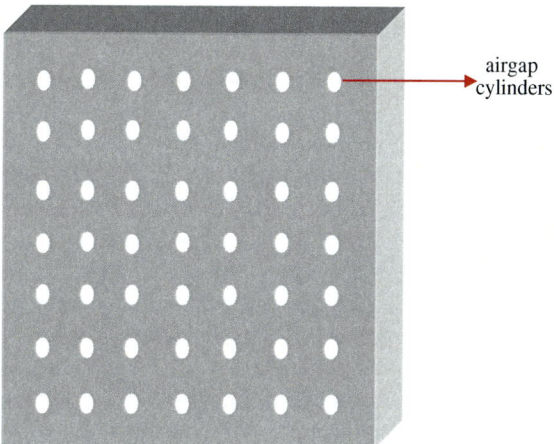

airgap
cylinders

photonic substrate are thick. Sharma et al. (2008) also determined that when the width of microstrip is 36 μm, the performance of antenna is good with low return loss.

RU Yu-xing et al. (2011) designed a patch antenna for an operating frequency of 12.5 GHz using photonic crystal (PC) substrate and compared the results obtained with the results of a conventional patch antenna and reported that the antenna with a PC substrate could achieve high gain and high directivity than a conventional patch antenna. The obtained gain using the designed patch antenna with PC substrate is 5 dB more than the gain obtained using a conventional patch antenna.

Tiwari et al. (2011) presented a design of rectangular microstrip patch antenna using Photonic band gap crystal that can be used for 60 GHz communications. The reported directivity, bandwidth, and radiation efficiency of this design at a frequency of 60 GHz are 9.143 dBi, 20.53 %, and 0.7674, respectively.

Singh et al. (2015) presented a design of trapezoidal microstrip patch antenna that can be used for high speed applications in THz frequency band ranging from 0.88 to 1.62 THz. The dielectric substrate designed consists of photonic crystals to reduce the loss due to surface waves. It was reported that the designed antenna has better gain, VSWR, and return loss compared to conventional antenna. The reported design also exhibits a wider bandwidth ranging from 1.2 to 1.62 THz.

Photonic band gap substrate emerged as a promising solution for improving the performance characteristics of microstrip patch antenna. Hence in the next section, a microstrip patch antenna was designed using PBG substrate, and the improvement in performance characteristics was analyzed.

5.2.2 PBG based THz Patch Antenna

The dimensions of microstrip patch antenna with PBG substrate are shown in Fig. 25.

The PBG substrate is designed by implanting air gap cylinders of radius 7.5 μm on a substrate. The center-to-center distance between each air gap is 100 μm.

The proposed model is designed and simulated in FEM-based software using MATLAB interfacing. The designed rectangular microstrip patch antenna with PBG substrate gives a return loss of −52.42 dB at a resonating frequency of 0.778 THz (Figs. 26, 27, 28, 29, and 30). The obtained gain and directivity at 0.778 THz are 13.58 dB and 13.82 dBi, respectively (Table 4).

It is observed that the return loss, gain, and directivity using a PBG substrate are improved when compared to a conventional microstrip patch antenna without PBG substrate of same dimensions. The corresponding 3D radiation pattern and the polar plot of the antenna are given in Figs. 27, 28, 29, and 30.

5.2.3 Optimization PBG Patch Antenna

As mentioned in the previous section, the surface wave loss can be reduced by the use of photonic band gap (PBG) substrate, which is obtained by implanting air gap cylinders periodically on a substrate. As the PBG substrate design affects the reduction of surface wave loss, it is very important to get an optimized design.

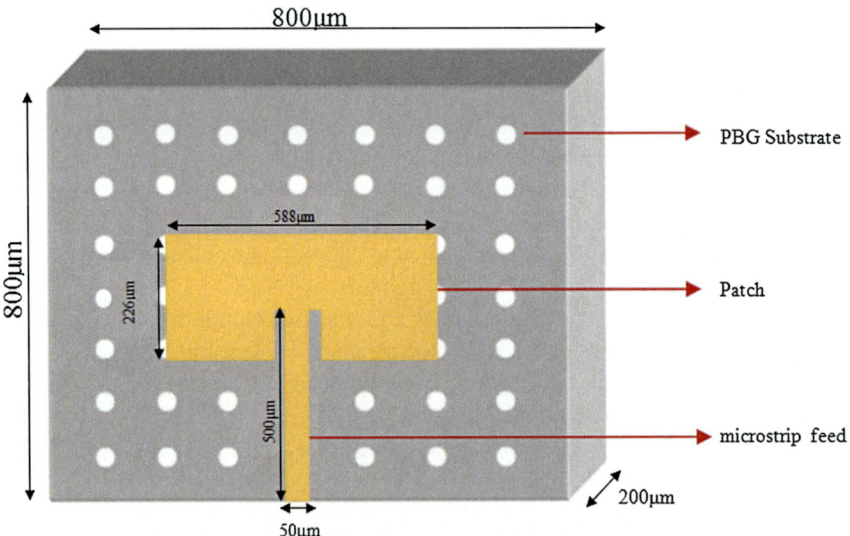

Fig. 25 Rectangular microstrip patch antenna with PBG substrate

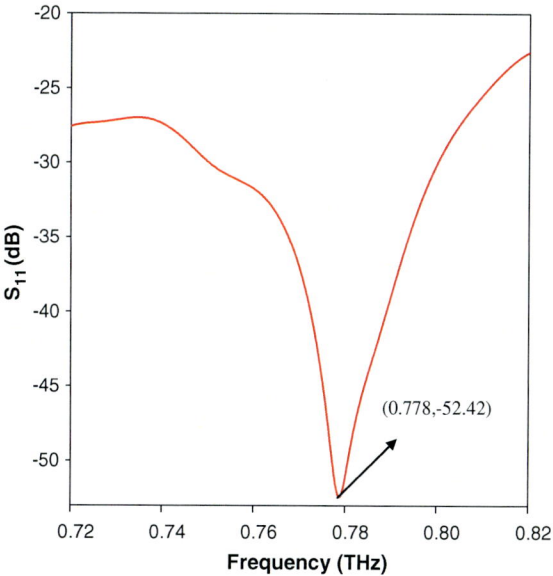

Fig. 26 Return loss characteristics of PBG based microstrip patch antenna

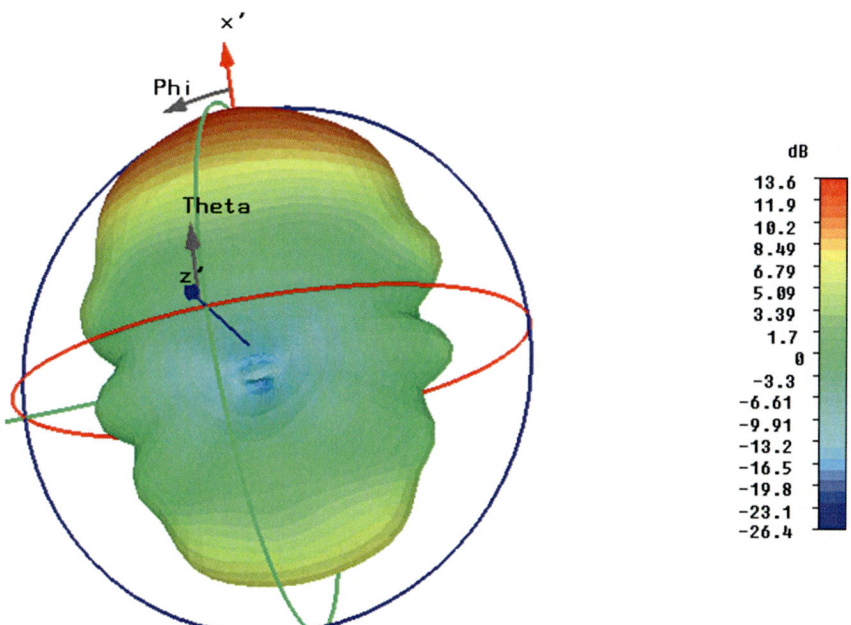

Fig. 27 3-D radiation pattern (Gain) of PBG based patch antenna

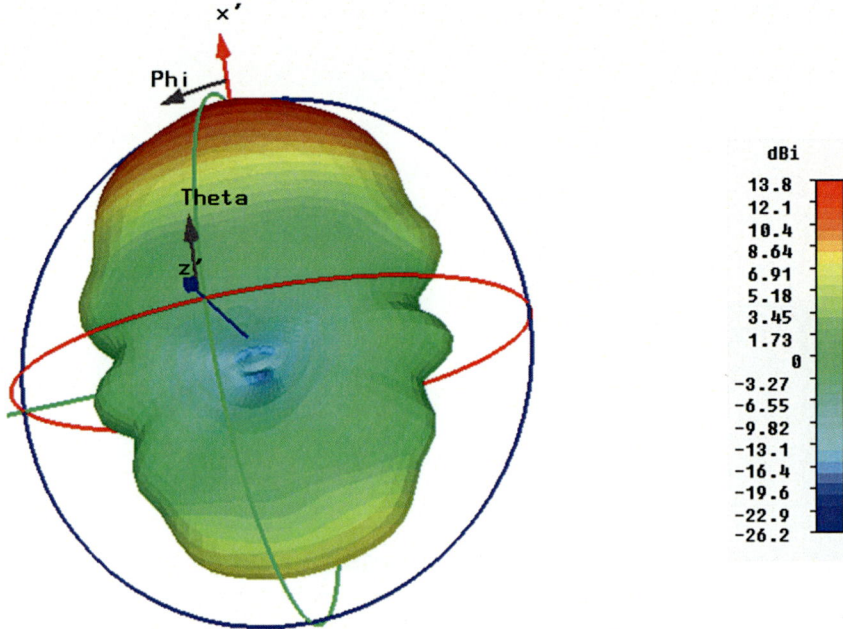

Fig. 28 3-D radiation pattern (directivity) of PBG based patch antenna

Fig. 29 Polar plot (phi = 90) of PBG based patch antenna

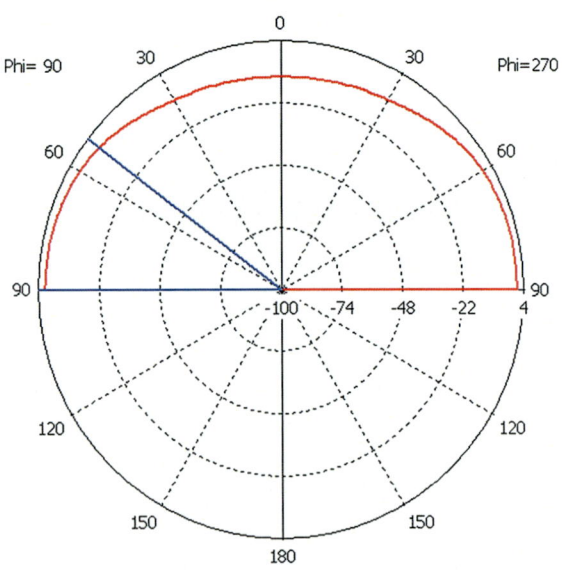

Theta / Degree vs. dBi

Fig. 30 Polar plot (phi = 0) of PBG based patch antenna

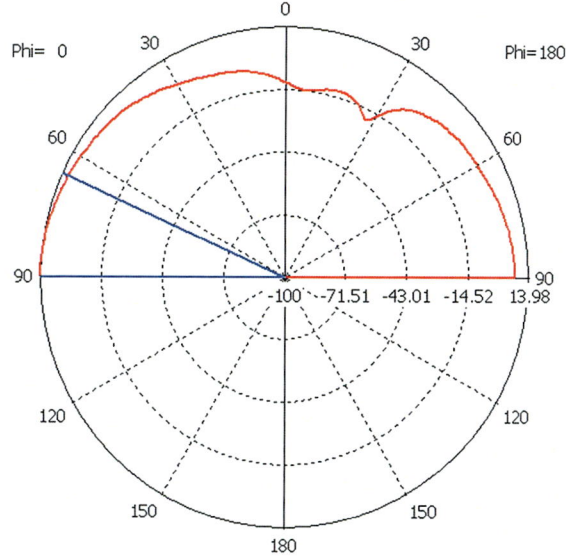

Theta / Degree vs. dBi

Table 4 Performance parameters of PBG based patch antenna

Frequency	0.778 THz
Return loss	−52.42 dB
Gain	13.58 dB
Directivity	13.82 dBi

Soft computing techniques (Appendix C) are best suited for these kinds of design optimization problems where multiple parameters have to be optimized. Multiobjective particle swarm optimization (Appendix C) is used here for design optimization of the PBG based antenna where both the periodicity of the air gap cylinders of the PBG substrate, diameter of the cylinder, and the thickness of the PBG substrate are varied to achieve a high gain antenna.

The optimized thickness of the PBG substrate is 300 μm. Thus the dimensions of the substrate used in this optimized design are $800 \times 800 \times 300 \ \mu m^3$ and is made of a material with a dielectric constant of 9.1. The dimensions of rectangular patch designed are $588 \times 226 \ \mu m^2$. A microstrip line of width 149.89 μm is used to feed the radiating patch. The width of the patch is calculated in order to provide a characteristic impedance of 50 Ω. The thickness of patch and microstrip feed is 36 μm. The feed inset position is at a distance of 88.67 μm from the lower end of the patch in order to obtain an impedance matching of 50 Ω. The PBG substrate is designed by implanting air gap cylinders of radius 7.5 μm periodically on a substrate. The center-to-center distance of 100 μm is maintained between each air gap. The optimized PBG based patch antenna is as shown in Fig. 31.

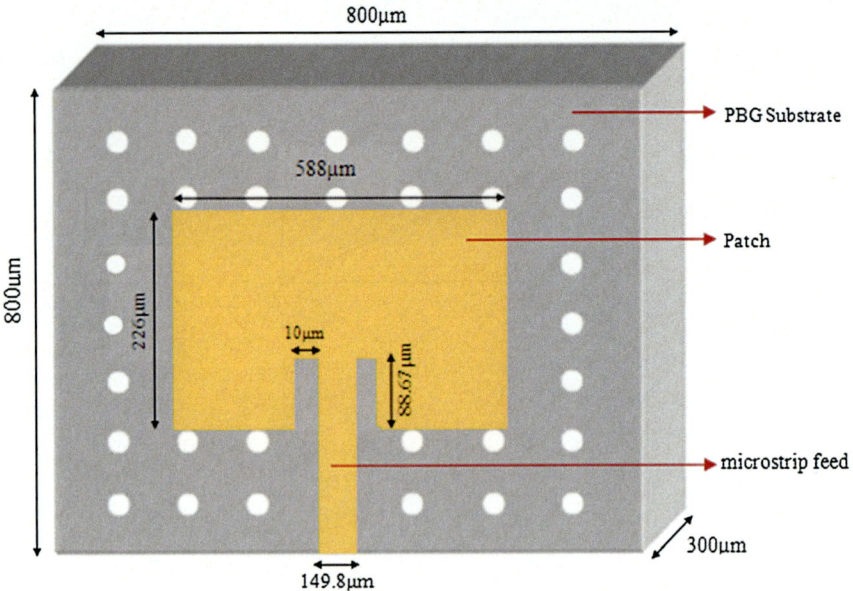

Fig. 31 PBG microstrip patch antenna with increased thickness

Fig. 32 Return loss
characteristics of optimized
PBG patch antenna

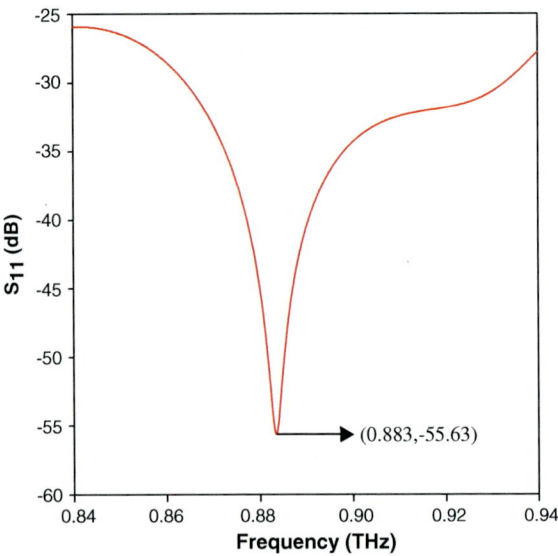

The performance improvement of the PBG based patch antenna with increased thickness can be observed from Figs. 32, 33, 34, 35 and 36. The return loss, gain, and directivity of the proposed model are −55.63 dB, 17.99 dB, and 18.22 dBi, respectively, (Table 5) which shows an improvement in the performance of an antenna.

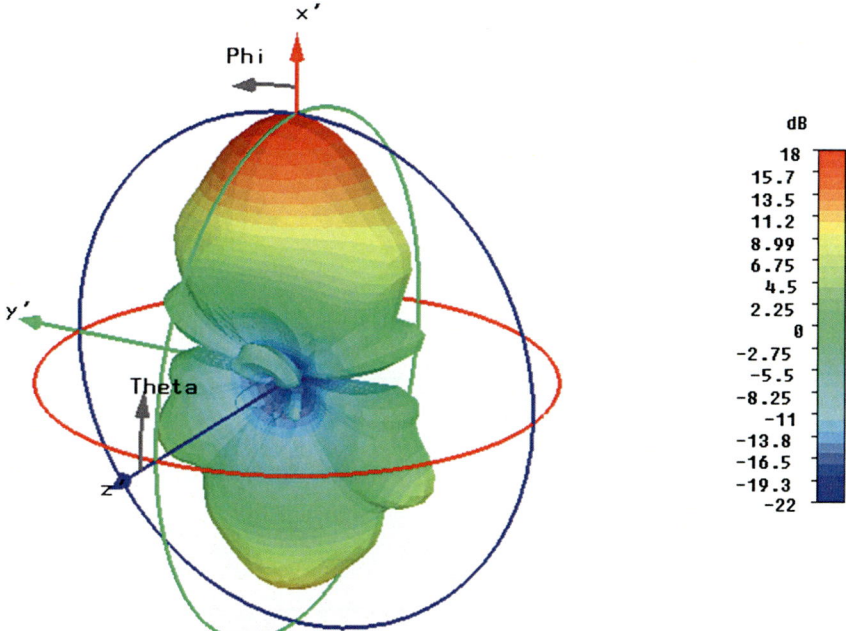

Fig. 33 3-D radiation pattern (gain) of optimized PBG patch antenna

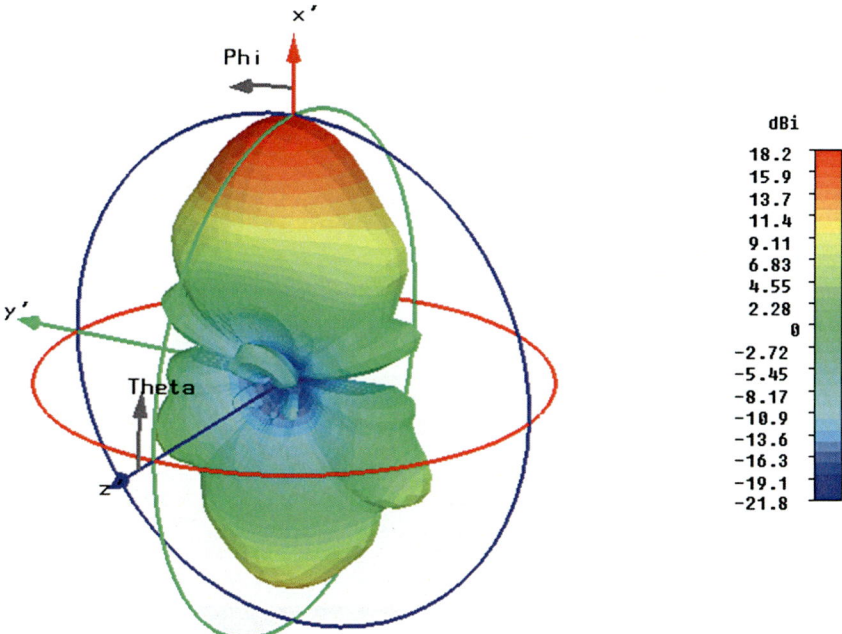

Fig. 34 3-D radiation pattern (directivity) of optimized PBG patch antenna

The 3D radiation pattern of the PBG based optimized antenna is shown in Fig. 33. The simulation results show that the gain of the antenna is 17.99 dB and directivity is 18.22 dB (Fig. 34), which is better than the widely used communication antennas. The corresponding polar plots at principle plane are shown in Figs. 35 and 36.

Fig. 35 Polar plot (phi = 90) of optimized PBG based patch antenna

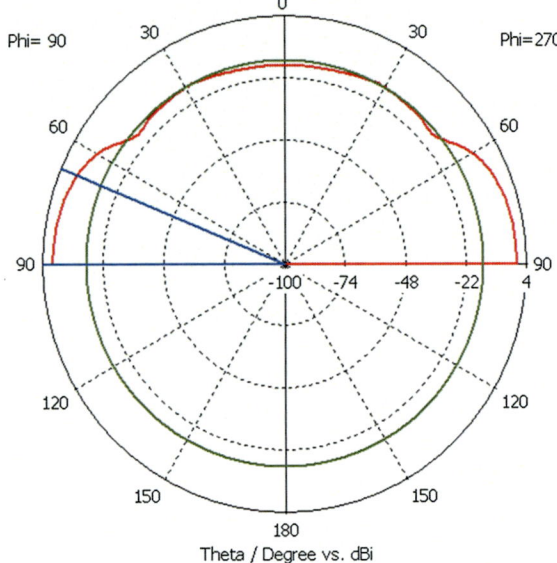

Fig. 36 Polar plot (phi = 0) of optimized PBG based patch antenna

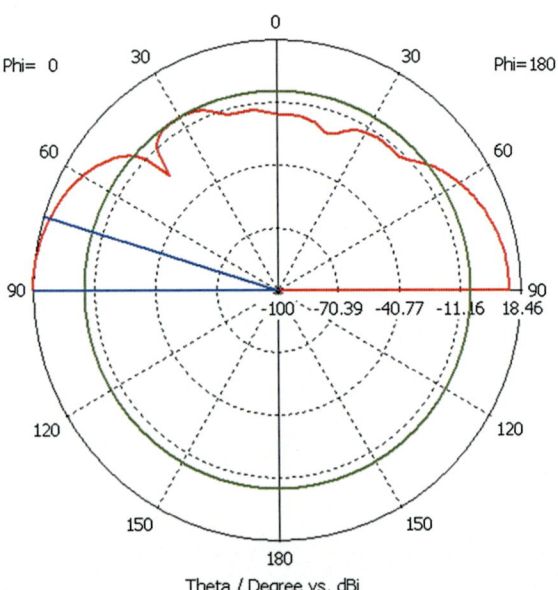

Table 5 Performance parameters of optimized PBG patch antenna

Frequency	0.883 THz
Return loss	−55.63 dB
Gain	17.99 dB
Directivity	18.22 dBi

6 Conclusion

In this work, possibility of terahertz spectrum in wireless space communication has been explored keeping in view of the increasing demand in high data rate transmission. It has been observed that the required data rate may increase to tens of Gbits/s in near future.

An effort has been made in this work to design a terahertz antenna with high gain and high directivity to achieve high data rate wireless communication. Microstrip patch antenna is designed because of its several advantages such as compact size, portable and conformable nature, low cost, ease of installation and fabrication, suitability for arrays and its performance characteristics. Toward enhancement of the terahertz antenna gain, an optimized microstrip antenna with PBG substrate has been designed. The PBG substrate reduces the surface wave loss, hence increases the gain as well as directivity. The feed is inserted into the patch so as to obtain an impedance matching between patch and feed line of characteristic impedance of the order of 50 Ω. The antenna model is designed and simulated using a FEM-based software and has been optimized using PSO-based computational engine.

The obtained values of return loss, gain, and directivity of the designed model are −55.63 dB, 17.99 dB, and 18.22 dBi, respectively, which show an improvement as compared to those of a conventional patch antenna without PBG substrate. Hence the designed antenna can be used for various wireless applications such as aircraft collision avoidance system, global positioning systems, telemetry, on-vehicle satellite links, missile radars, interorbital communications, communications inside space vehicle, and other space-based applications because of its compact size and enhanced performance characteristics.

References

Azarbar, A., M.S. Masouleh, and A.K. Behbahani. 2014. A new terahertz microstrip rectangular patch array antenna. *International Journal of Electromagnetics and Applications* 4(1): 25–29.

Al-Tarifi, M.A., D.E. Anagnostou, A.K. Amert, and K.W. Whites. 2015. The puck antenna: a compact design with wideband, high-gain operation. *IEEE Transactions on Antennas and Propagation* 63(4): 1868–1872.

Babanrao, P.S., S.P. Bhosale, and A.Y. Kazi, 2014. Enhancement of return loss and efficiency of microstrip slotted patch antenna using line shape defected ground structure. *International Journal on Recent and Innovation Trends in Computing and Communication* 2: 343–346. ISSN: 2321–8169.

Cai, Y., I. Brener, J. Lopata, J. Wynn, L. Pfeiffer, and J. Federici. 1997. Design and performance of singular electric field terahertz photoconducting antennas. *Applied Physics Letters* 71(15): 2076–2078.

Cherednichenko, S., M. Kroug, H. Merkel, P. Khosropanah, A. Adam, E. Kollberg, D. Loudkov, G. Gol'tsman, B. Voronov, H. Richter, and H.W. Huebers. 2002. 1.6 THz heterodyne receiver for the far infrared space telescope *Physica C: Superconductivity* 372–376: 427–431.

Chouvaev, D., L. Kuzmin, M. Tarasov, P. Sundquist, M. Willander, and T. Claeson. 1998. Normal metal hot-electron microbolometer with Andreev mirrors for THz space applications. In *Proceedings of the 9th International Symposium on Space Terahertz Technology*, Pasadena, pp. 331–335.

Ezdi, K., B. Heinen, C. Jordens, N. Vieweg, N. Krumbholz, R. Wilk, M. Mikulics, and M. Koch. 2009. A hybrid time-domain model for pulsed terahertz dipole antennas. *Journal of the European Optical Society—Rapid Publications* 4: 09001(1)–09001(7).

Federici, J., and L. Moeller. 2010. Review of terahertz and subterahertz wireless communications. *Journal of Applied Physics* 107: 111101(1)–111101(22).

Fitch, M.J., and R. Osiander. 2004. Terahertz waves for communications and sensing. *Johns Hopkins APL Technical Digest* 25: 348–355.

Hagmann, C., D.J. Benford, A.C. Clapp, P.L. Richards, and P. Timbiet. 1992. A broadband THz receiver for low background space applications. In *Third International Symposium on Space Terahertz Technology* 678–687.

Han, H.P., J.G. Yuan, and J.C. Tong. 2015. Design of THz space application system. *Journal of Computer and Communications* 3: 61–65.

Hwu, S.U., K.B. deSilva, and C.T. Jih. 2013. Terahertz (THz) wireless systems for space applications. *Sensors Application Symposium*, pp. 171–175, Feb 2013.

Jepsen, P.U., R.H. Jacobsen, and S.R. Keiding. 1996. Generation and detection of terahertz pulses from biased semiconductor antennas. *Journal of the Optical Society of America* 13(11): 2424–2436.

Jha, K.R., and G. Singh. 2010. Analysis and design of enhanced directivity microstrip antenna at terahertz frequency by using electromagnetic bandgap material. *International Journal of Numerical Modelling: Electronic Networks, Devices and Fields* 410–424 doi:10.1002/jnm.787.

Karasik, B.S., A.V. Sergeyev, D. Olaya, J. Wei, M.E. Gershenson, J.H. Kawamura, and W.R. McGrath. 2005. A photon counting hot-electron bolometer for space THz spectroscopy. In *16th International Symposium on Space Terahertz Technology*.

Karpov, A., D. Miller, F. Rice, J. Zmuidzinas, J.A. Stern, B. Bumble, and H.G. LeDuc. 2003. Low noise SIS mixer for the band 1.1–1.25 THz of the Herschel space radio telescope. In *14th International Symposium on Space Terahertz Technology*, 1 pp.

Kumar, A., J. Kaur, and R. Singh. 2013. Performance analysis of different feeding technique. *International Journal of Emerging Technology and Advanced Engineering* 3(3).

Li D., and Yi Huang. 2006. Comparison of terahertz antennas. In *Proceedings of First European Conference on Antennas and Propagation*, pp. 1–5, Nov 2006.

Llombart, N., K.B. Cooper, R.J. Dengler, T. Bryllert, and P.H. Siegel. 2010. Confocal ellipsoidal reflector system for a mechanically scanned active terahertz imager. *IEEE Transactions on Antennas and Propagation* 58(6): 1834–1841.

Maiwald, F., J.C. Pearson, J.S. Ward, E. Schlecht, G. Chattopadhyay, J. Gill, R. Ferber, R. Tsang, R. Lin, A. Peralta, B. Finamore, W. Chun, J.J. Baker, R.J.Dengler, H. Javadi, P. Siegel and 1. Mehdi. 2004. Solid-state terahertz sources for space applications. In *Joint 29th International Conference on Infrared and Millimeter Waves and 12' International Conference on Terahertz Electronics*, pp. 767–768.

Piao, Z., M. Tani, and K. Sakai. 2000. Carrier dynamics and terahertz radiation in photoconductive antennas. *Japanese Journal of Applied Physics* 39: 96–100.

Pitra, K., Z. Raida, and H. L. Hartnagel, 2013. Design of circularly polarized terahertz antenna with interdigital electrode photomixer. In *7th European Conference on Antennas and Propagation*, pp. 2431–2434.

Sharma, A., V.K. Dwivedi, and G. Singh. 2008. THz rectangular patch microstrip antenna design using photonic crystal as substrate. In *Proceedings of Progress in Electromagnetics Research Symposium*, pp. 161–165.

Sharma, A., V.K. Dwivedi, and G. Singh. 2009. THz rectangular microstrip patch antenna on multilayered substrate for advance wireless communication systems. In *Proceedings of Progress in Electromagnetics Research Symposium*, pp. 627–631.

Siegel, P.H., 2010. THz for space: The golden age. In *IEEE MTT-S International Microwave Symposium Digest*, pp. 816–819.

Singh, R., C. Rockstuhl, C. Menzel, T.P. Meyrath, M. He, H. Giessen, F. Lederer, and W. Zhang. 2009. Spiral-type terahertz antennas and the manifestation of the Mushiake principle. *Opt. Express* 17(12): 9971–9980.

Singh, A., and S. Singh. 2015. A trapezoidal microstrip patch antenna on photonic crystal substrate for high speed THz applications. *Photonics and Nanostructures—Fundamentals and Applications* 14: 52–62.

Smith, D.R., J.B. Pendry, and M.C.K. Wiltshire. 2004. Metamaterials and negative refractive index. *Science* 305: 788–792.

Sobis, P., V. Drakinskiy, A. Emrich, H. Zhao, T. Bryllert, A.Y. Tang, J. Hanning, and J. Stake. 2013. 300 GHz to 1.2 THz GaAs Schottky membrane TMIC's for next generation space missions. In *24th International Symposium on Space Terahertz Techniques*, 1pp.

Tiwari, R. N., P. Kumar, and N. Bisht. 2011. Rectangular microstrip patch antenna with photonic band gap crystal for 60 GHz communications. In *Proceedings of Progress in Electromagnetic Research Symposium*, pp. 856–859, Sept 2011.

Tyagi, S., and K. Vyas. 2013. Bandwidth enhancement using slotted U-shape microstrip patch antenna with PBG ground. *International Journal of Advanced Technology and Engineering Research (IJATER)* 3: 23–27.

Yu-xing, R., S. Jiang-dong, W. Ge, G. Bo, and T. Xiao-jian. 2011. Design of a rectangular patch antenna with a photonic crystal substrate. *The Journal of China Universities of Posts and Telecommunications* 2 (Supplement): 161–163.

Zhang, J., Y. Hong, S. Braunstein, and K. Shore. 2004. Terahertz pulse generation and detection with LT-GaAs photoconductive antenna. *IEE Proceedings: Optoelectronics* 151(2): 98–101.

Zimmermann, R., T. Rose, T.W. Crowe, and TW. Grein. 1995. An all-solid-state 1 THz radiometer for space applications. In *Sixth International Symposium on Space Terahertz Technology*, pp. 13–27.

Appendix A
Models of PBG based Microstrip Patch Antenna

Various dimensional models of rectangular microstrip patch antenna with PBG substrate are designed and simulated using FEM-based software, and the results are tabulated in Table A.1.

Table A.1. Performance characteristics of various microstrip patch antenna models using PBG substrate

S. no	Design parameters		Gain (dB)	Directivity (dBi)	Return loss (dB)	Resonating frequency (THz)
1	Patch dimensions	$588 \times 226 \times 36 \ \mu m^3$	13.25	13.58	−66.27	0.733
	Substrate dimensions	$900 \times 900 \times 200 \ \mu m^3$				
	Substrate ε	9.1				
	Air gap radius	7.5 μm				
	Air gap periodicity	200 μm				
	Feed width	99.89 μm				
	Inset feed position	80 μm				
2	Patch dimensions	$588 \times 226 \times 36 \ \mu m^3$	13.35	13.70	−28.14	0.7464
	Substrate dimensions	$900 \times 900 \times 200 \ \mu m^3$				
	Substrate ε	9.1				
	Air gap radius	10 μm				
	Air gap periodicity	200 μm				
	Feed width	99.89 μm				
	Inset feed position	80 μm				

(continued)

© The Author(s) 2016

B. Choudhury et al., *PBG based Terahertz Antenna for Aerospace Applications*,
SpringerBriefs in Computational Electromagnetics,
DOI 10.1007/978-981-287-802-1

Table A.1. (continued)

S. no	Design parameters		Gain (dB)	Directivity (dBi)	Return loss (dB)	Resonating frequency (THz)
3	Patch dimensions	$588 \times 226 \times 36^3$	13.66	13.94	−31.02	0.7
	Substrate dimensions	$800 \times 800 \times 200 \ \mu m^3$				
	Substrate ε	9.1				
	Air gap radius	7.5 μm				
	Air gap periodicity	200 μm				
	Feed width	99.89 μm				
	Inset feed position	80 μm				
4	Patch dimensions	$588 \times 226 \times 36 \ \mu m^3$	13.45	13.74	−26.89	0.7
	Substrate dimensions	$800 \times 800 \times 200 \ \mu m^3$				
	Substrate ε	9.1				
	Air gap radius	10 μm				
	Air gap periodicity	200 μm				
	Feed width	99.89 μm				
	Inset feed position	80 μm				
5	Patch dimensions	$588 \times 226 \times 36 \ \mu m^3$	12.63	12.98	−37.255	0.744
	Substrate dimensions	$900 \times 900 \times 200 \ \mu m^3$				
	Substrate ε	10.2				
	Air gap radius	10 μm				
	Air gap periodicity	200 μm				
	Feed width	99.89 μm				
	Inset feed position	80 μm				
6	Patch dimensions	$588 \times 226 \times 36 \ \mu m^3$	11.74	12.14	−40.293	0.738
	Substrate dimensions	$900 \times 900 \times 200 \ \mu m^3$				
	Substrate ε	10.2				
	Air gap radius	7.5 μm				
	Air gap periodicity	200 μm				
	Feed width	99.89 μm				
	Inset feed position	80 μm				

(continued)

Table A.1. (continued)

S. no	Design parameters		Gain (dB)	Directivity (dBi)	Return loss (dB)	Resonating frequency (THz)
7	Patch dimensions	$588 \times 226 \times 36 \ \mu m^3$	11.66	11.94	−61.577	0.66
	Substrate dimensions	$800 \times 800 \times 200 \ \mu m^3$				
	Substrate ε	10.2				
	Air gap radius	7.5 μm				
	Air gap periodicity	200 μm				
	Feed width	99.89 μm				
	Inset feed position	80 μm				
8	Patch dimensions	$588 \times 226 \times 36 \ \mu m^3$	11.99	12.30	−33.23	0.7
	Substrate dimensions	$800 \times 800 \times 200 \ \mu m^3$				
	Substrate ε	10.2				
	Air gap radius	10 μm				
	Air gap periodicity	200 μm				
	Feed width	99.89 μm				
	Inset feed position	80 μm				
9	Patch dimensions	$588 \times 226 \times 36 \ \mu m^3$	14.11	13.68	−38.19	0.8
	Substrate dimensions	$800 \times 800 \times 200 \ \mu m^3$				
	Substrate ε	9.1				
	Air gap radius	10 μm				
	Air gap periodicity	100 μm				
	Feed width	99.89 μm				
	Inset feed position	80 μm				
10	Patch dimensions	$588 \times 226 \times 36 \ \mu m^3$	16.21	16.58	−36.48	0.7036
	Substrate dimensions	$800 \times 800 \times 300 \ \mu m^3$				
	Substrate ε	9.1				
	Air gap radius	10 μm				
	Air gap periodicity	100 μm				
	Feed width	149.8 μm				
	Inset feed position	88.67 μm				

(continued)

Table A.1. (continued)

S. no	Design parameters		Gain (dB)	Directivity (dBi)	Return loss (dB)	Resonating frequency (THz)
11	Patch dimensions	$588 \times 226 \times 36 \ \mu m^3$	13.73	14.08	−46.98	0.66
	Substrate dimensions	$800 \times 800 \times 300 \ \mu m^3$				
	Substrate ε	10.2				
	Air gap radius	10 μm				
	Air gap periodicity	100 μm				
	Feed width	149.8 μm				
	Inset feed position	88.67 μm				
12	Patch dimensions	$588 \times 226 \times 36 \ \mu m^3$	13.70	14.08	−65.56	0.656
	Substrate dimensions	$800 \times 800 \times 300 \ \mu m^3$				
	Substrate ε	10.2				
	Air gap radius	7.5 μm				
	Air gap periodicity	100 μm				
	Feed width	149.8 μm				
	Inset feed position	88.67 μm				

Appendix B
Feeding Types for Microstrip Patch Antennas

Recent progressions in space wireless communication has resulted in the growth of usage of microstrip patch antennas because of its small size, lower cost, ease of fabrication and installation, suitability for arrays, and radiation characteristics which proved them to be advantageous over other conventional antennas.

The feeding of RF power between the microstrip line and radiating patch can be done in two methods. One method is to feed directly using a connecting element. The feeding techniques that are widely used in this method are

- Microstrip Line Feed
- Coaxial Line Feed

And the other method is to use electromagnetic field coupling. There are two widely used feeding techniques to transfer RF power in this method. They are

- Proximity Coupling
- Aperture Coupling

In microstrip feeding technique, the feed is connected directly to the edge of the patch. But this results in poor impedance matching. An inset cut feed where the microstrip line inserted into the patch by making an inset cut provides proper matching which can be obtained controlling the inset position. The main advantage of using microstrip line feeding technique is its ease of fabrication. The disadvantage of this feeding technique is its limited bandwidth.

In coaxial feed, the inner conductor of coaxial connector is soldered to the radiating patch through the dielectric, and the outer conductor is joined to the ground plane. Impedance matching between the feed and the patch can be obtained by placing the feed at a proper location inside the patch. Advantage of this type of feeding is that the feed can be placed at any desirable location on patch in order to obtain proper impedance matching. The disadvantage of this feeding technique is its small bandwidth and difficulty in fabrication because a hole has to be drilled on the substrate. And it is difficult to obtain proper impedance matching when a thicker substrate is used for antenna design.

© The Author(s) 2016
B. Choudhury et al., *PBG based Terahertz Antenna for Aerospace Applications*,
SpringerBriefs in Computational Electromagnetics,
DOI 10.1007/978-981-287-802-1

In aperture couple feed, an aperture is made on the ground plane through which the coupling of energy between the patch and the feed line takes place. There will be no physical connection between the feed and the radiating patch. The advantage of this feeding technique is that it provides higher bandwidth. The major disadvantage of this feeding technique is its difficulty in fabrication and proper impedance matching requires careful alignment.

In proximity coupled feeding technique, the feed is placed between two layers of dielectric substrate and the patch is placed over the upper substrate. The coupling mechanism between the radiating patch and the feed is identical to a capacitor. The major advantage of this feeding technique is that it has less spurious radiation and more bandwidth. But its fabrication is difficult.

The properties of the antenna can be varied by varying the design parameters of antenna. The resonant frequency of the antenna can be controlled by varying the length of the patch. The radiation pattern and the input impedance of the antenna can be controlled by varying the width of the patch. The bandwidth of the antenna can be increased by increasing the height of the substrate. Although the use of thick substrates may result in surface wave trapping which can be avoided using a PBG substrate that has air gap cylinders periodically entrenched on the substrate.

Appendix C
Multiobjective Particle Swarm Optimization

Particle swarm optimization (PSO) is a soft-computing technique that emulates swarm intelligence. Kennedy and Eberhart first proposed this algorithm in 1995. The basic functioning of this algorithm can be best explained using the example of a swarm of bees that are trying to locate a point in a field that has maximum density of flowers. Initially, all the bees randomly disperse throughout the field as they have no knowledge about the position of the point that has maximum density of flowers. Now each bee finds out the density of flowers at each location, informs the other bees, and determines the direction that has the highest density of flowers. In this way, each bee moves in a direction by comparing its own best value and the best value in the swarm. Now all the bees move in that particular direction and finds out the exact point in that location that has the highest density of flowers. The behavior of these bees is formulated in a computer algorithm. The "density of flowers" is equivalent to a problem specific objective function. The final point with maximum flower density is equivalent to the one which results in minimum value of the objective function. The basic terminology that is used in particle swarm optimization is as follows:

Particle	Potential solution in the solution space (analogous to a bee)
Fitness	Result of evaluation of the problem's objective function for the particles co-ordinates
Personal Best (pbest)	Best value of fitness obtained for each particle
Global Best (gbest)	Best value of fitness among all particles in all iterations

© The Author(s) 2016
B. Choudhury et al., *PBG based Terahertz Antenna for Aerospace Applications*,
SpringerBriefs in Computational Electromagnetics,
DOI 10.1007/978-981-287-802-1

Algorithm for PSO: The algorithm implemented for PSO is as follows:

Step 1   *First initialize the random positions and velocities for each particle*
           A random position within the N dimensional solution space [xMin, xMax]
           is assigned to each and every particle. Here it is assumed that each
           dimension is of the same range. However, this may not be the same for
           some problems and hence, should be assigned suitably. Then each particle
           is assigned a velocity lying within the range [vMin, vMax]. The values for
           vMin and vMax can be calculated as given below [15]

$$v\text{Min} = -0.2(x\text{Max} - x\text{Min})$$
$$v\text{Max} = 0.2(x\text{Max} - x\text{Min}) \tag{C.1}$$

Step 2   *Evaluate the fitness and determine* pbest, gbest The fitness function
           (objective function) f for the current position of each and every particle is
           determined. Initially, these evaluated values are considered as pbest. The
           calculation of fitness function of each particle is repeated for all the
           subsequent iterations. If the calculated fitness value is less than the current
           pbest, then the calculated fitness is considers as the new pbest. The
           minimum of all pbest is considered as the gbest and is updated when a
           fitness value that is lesser than the current gbest value is obtained.

Step 3   *Update Position and Velocity* Depending on the values of pbest and gbest,
           the position (x) and velocity (v) for each particle is updated using the
           following given equations:

$$v = w \times v + c_1 r_1(\text{pbest} - x) + c_2 r_2(\text{gbest} - x) \tag{C.2}$$

$$x = x + v \tag{C.3}$$

           where w is called the inertial constant, c_1 is called the social constant, and
           c_2 is called cognitive constant. From Eqs. (C.2) and (C.3), it can be
           observed that the movement of particles is described using Newtonian
           mechanics. For best results, the value of w is considered as 1 at the start of
           the algorithm and decremented 0.4 at the end of the algorithm. Further,
           research has found that the rate of convergence in the algorithm can be
           increased using a value of 0.5 for the social and cognitive parameters.

Step 4   *Run loop* Go to Step 2 and carry out each step for N_t times.

The above described algorithm helps in finding solutions to the problems in
which the output is determined by a single fitness function. But in some real cases,
the output might depend on more than one fitness function. In such cases, the aim of

the PSO is to determine the points in the solution space in which the fitness of one function can be enhanced only by degrading the fitness's of the other functions. All such points obtained are called as *pareto front*. Examples of some frequently obtained *pareto fronts* are then given through solution to various fitness functions.

MOPSO algorithm is also similar to that of the PSO algorithm discussed in the earlier section. But, the only difference between PSO and MOPSO is that, in MOPSO each and every particle in the solution space has its own gbest unlike PSO, in which a single value is assigned as global best (gbest).

When the algorithm runs, all dominated solutions are included to '*archive*'(a special matrix). The archive is updated for every step in the iteration in order to collect a set of solutions that are nondominated with respect to each other. The gbest of each particle which is the nearest dominated solution in terms of distance is used in the velocity update equation given in Eqs. (C.2) and (C.3). The pbest of each particle is the best dominated solution visited by that particle. The output of the MOPSO algorithm is represented by the values in the archive at the end of the iteration. These values represent points where the fitness of one function cannot be enhanced unless degrading the fitness of the other functions. The user has the choice of choosing whichever value from the archive depending upon the application and the significance that was assigned to each fitness function.

Testing of Multiobjective PSO

MOPSO code can be obtained by making little modifications to the single objective PSO code (Jin and Nanbo 2008). *Pareto fronts* for five different test functions were obtained and the robustness of this MOPSO code was tested.

Case 1: Pareto front of convex-uniform type
The following functions were considered as the fitness functions. Figure C.1 represents the *pareto front* that was obtained when MOPSO was applied to optimize both these functions at the same time.

$$f_1 = \frac{1}{N} \sum_{i=1}^{N} x_i^2 \qquad (C.4)$$

$$f_2 = \frac{1}{N} \sum_{i=1}^{N} (x_i - 2)^2 \qquad (C.5)$$

Case 2: Pareto front of concave type
The following functions were considered as the fitness functions. Figure C.2 shows the *pareto front* that was obtained when MOPSO was applied to optimize both these functions at the same time.

Figure C.1 Convex, uniform *pareto front* obtained by optimizing Eqs. (C.4) and (C.5)

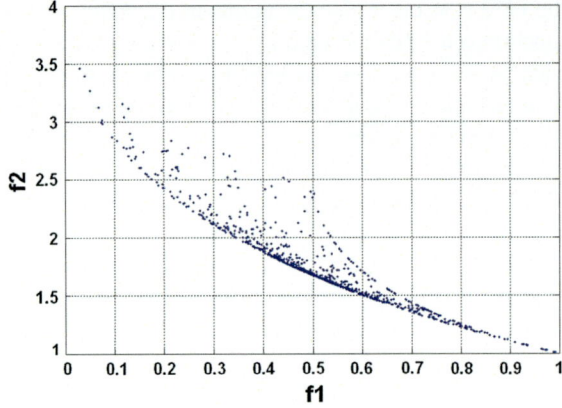

Figure C.2 *Pareto front* (Concave) obtained by optimizing Eqs. (C.6) and (C.7)

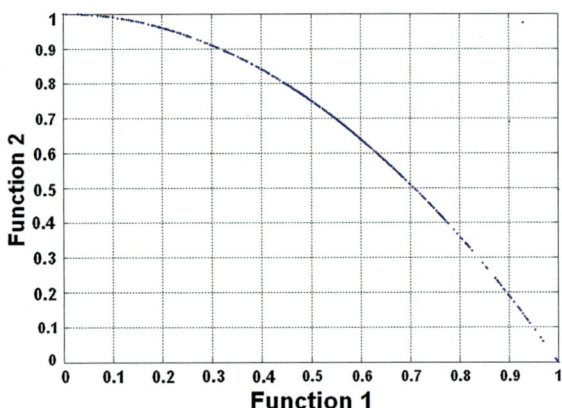

$$f_1 = x_1 \tag{C.6}$$

$$f_2 = g \times \left(1 - \left(\frac{f_1}{g}\right)^2\right) \tag{C.7}$$

$$g = 1 + \frac{9}{N-1}\sum_{i=1}^{N} x_i \tag{C.8}$$

Reference

Jin and Nambo. 2008. *Particle Swarm Optimization in Engineering Electromagnetics*, UMI Dissertion Services.

About the Book

This book focuses on high-gain antennas in the terahertz spectrum and their optimization. The terahertz spectrum is an unallocated EM spectrum, which is being explored for a number of applications, especially to meet increasing demands of high data rates for wireless space communications. Space communication systems using the terahertz spectrum can resolve the problems of limited bandwidth of present wireless communications without radio-frequency interference. This book describes design of such high-gain antennas and their performance enhancement using photonic band gap (PBG) substrates. Further, optimization of antenna models using evolutionary algorithm-based computational engine has been included. The optimized high-performance compact antenna may be used for various wireless applications, such as interorbital communications and on-vehicle satellite communications.

© The Author(s) 2016   47
B. Choudhury et al., *PBG based Terahertz Antenna for Aerospace Applications*,
SpringerBriefs in Computational Electromagnetics,
DOI 10.1007/978-981-287-802-1

Author Index

B. Choudhury et al., *PBG based Terahertz Antenna for Aerospace Applications*,
SpringerBriefs in Computational Electromagnetics,
DOI 10.1007/978-981-287-802-1

Subject Index

© The Author(s) 2016   51
B. Choudhury et al., *PBG based Terahertz Antenna for Aerospace Applications*,
SpringerBriefs in Computational Electromagnetics,
DOI 10.1007/978-981-287-802-1

Printed in the United States
By Bookmasters